BULK CARGO

A short introduction for loading, unloading and stowage of solid Bulk Cargoes including Draught Survey

CAPT. PETER GRUNAU

Table of Contents

1.1 What is Bulk Cargo? *4*
2.0 Preparation to take over the cargo 7
3.0 Important properties of the cargo 8
    3.1 Angle of Repose *11*
    3.2 Factors Affecting the Shear Strength that Determines the Angle of Repose .. *11*
    3.3 Influence of the Angle of Repose to Ships' Stability *12*
4.0 Loading and discharging Procedures 15
    4.1 Homogeneous Hold loading Condition (Fullyloaded) *19*
    4.2 Alternate Hold loading condition (Fullyloaded) *20*
    4.3 Block Hold loading and Part loaded condition *21*
    4.4 Load planning using a Load Master *24*
    4.5 On board loading Guidance *32*
    4.6 Ratio between Loading Rate and De-Ballast Rate *33*
    4.7 Loading program and de-ballasting program for loading of Bulk Carrier *35*
        4.7.1 Loading with two loaders at the same time *37*
5.0 Maintenance during the voyage 41
    5.1 Limiting Points of the Voyage *41*
    5.2 Voyage Planning *49*
6.0 Grain in Bulk 69
    6.1 Grain a sensitive cargo *69*
7.0 Stability Related Problems on Bulk Carrier 72
8.0 The Code of Safe Practice for Solid Bulk Cargoes 75
    8.1 General precautions *76*
    8.2 Bulk cargoes having an angle of repose less than or equal to 35 degrees *78*
    8.3 Bulk cargoes having an angle of repose greater than 35 degrees *78*
    8.4 Safety of personnel *79*
    8.5 Cargoes which may liquefy *80*

- 8.5.1 Correct course of action if loading Cargoes which may liquefy 81
- 8.6.1 Shifting of Mineral Concentrates and Ore in Bulk 85
- 8.7 Cargoes of Cargo Group A – Cargoes which may liquefy 85
  - *8.7.1 Cargoes which may shift related to their internal friction characteristics.. 85*
  - *8.7.2 Cargoes which might shift due to their moisture content 86*
- 8.7.3 Moisture content exceeding the safe transportable moisture limit 88
- 8.7.4 Where shifting of cargo is probable 89
  - *Consideration for the intact stability of the ship 89*
  - *8.7.5 Calculation of the cargo shift moments and the reduction in KG 90*
  - *8.7.5.1 Example Calculation to justify the necessity for this calculation........... 93*
- 8.7.6 Conclusion 96
  - *8.8 Direct Reduced Iron Ore 98*
    - 8.8.1 The Direct Reduction 99
    - 8.8.3 Unprocessed Ores 101
    - 8.8.4 Iron Ore Fines 102
    - 8.8.5 Sinter Feed 103
  - *8.9 Common Hazard of Bulk Cargo 104*
- 9.0 The Draught survey 110
  - *9.1 Reading the draughts 113*
    - 9.1.1 Measuring the draughts in a swell 113
    - 9.1.2 Draught reading on outboard side 114
  - *9.2 The common method to read the draught for the draught survey 114*
  - *9.3 The general requirements for a draught survey 115*
  - *9.4 Calculating the deductibles 116*
  - *9.5 Calculation Perpendicular corrections 117*
  - *9.6 The influence of the trim on the draught survey 119*
  - *9.7 Correction for deformation 120*
  - *9.8 Correct for heel and how to apply 122*
  - *9.9 The position of LCF 123*
  - *9.10 Trim Corrections 129*

9.10.1 1st Trim Correction.. 129

9.10.2 2nd Trim Correction... 130

9.11 The Density Correction 132

9.12 The constant - A variable in the calculation of the Draught Survey 133

9.13 Calculation of the displacement corrected.. 135

10.0 The Vetting Inspection 139

11.   Preparation of Hatches if loading Bulk Cargoes 140

    11.1   General preparation 140

    11.2   Hold Wash.. 146

    11.3 Disposal of wash water 147

    11.4 Bilge cleaning and preparation.. 147

12.0 Hatch cover maintenance 149

    12.1 Maintenance of Hatch Cover Structure... 150

    12.2 Testing of Hatch cover 151

    12.3 Maintenance and Repair 153

13.0 The Grain Code and the Intact Stability Requirements............................. 155

    13.1 Stability Regulations and Requirements when Loading Grain Products
.... 156

    13.2   Voids in Spaces Loaded with Grain 158

    13.3 The Loss of GZ if Grain Cargo Shifts.. 161

        13.3.1 Compensation for the Vertical Component of Shift of Grain............. 163

    13.3.3 The US Grain regulation and Requirement...................................... 169

    13.3.4 Calculation of the Residual Area 170

    13.3.5 Residual area according to the NCB... 183

14.0 Stowage of Grain 185

Bibliography 191

Table of Illustration.. 192

1.1 What is Bulk Cargo?

All kind of cargoes which have to be transported in large masses (Bulk), especially coal, iron ore, mineral products, alumina, grain[1], corn, milo etc. These cargoes will be shipped in special vessels- Bulk Carriers.

Each cargo has his own procedure for loading and discharging. The most important point for carrying bulk is the angle of repose. The angle of repose indicates the "Stability of the cargo". .The angle of repose will be given to the ship command either by the agent or charterer or can be found in the IMSBC Code (**I**nternational **M**aritime **S**olid **B**ulk **C**argoes Code) for Bulk Carrier.

The most widely recognised structural arrangement, which is identifying a bulk carrier is a single deck ship with double bottom, side tanks, single skin transverse framed side shell topside tanks and deck hatchway. The plating is supported by secondary stiffening members which transfer the loads to the primary structural members, like double bottom, web frames and girders. The cross deck and the double bottom structure are the main structural members, which provides the transverse strength of the ship to prevent against hull section distorting.

---

[1] For loading grain special regulation have to be followed

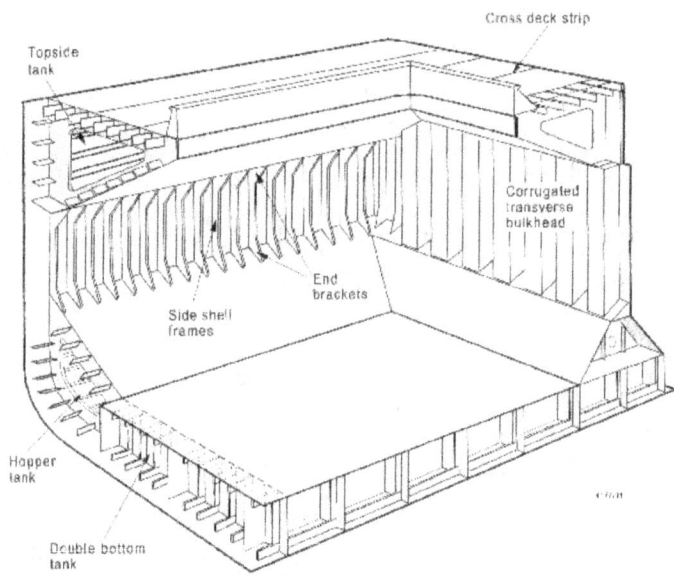

Illustration 1 Cargo Hold Structure of a single deck side skin Bulk carrier[1]

---

[1] /32Aus :IACS -Bulk carrier: Guidance and information on Bulk Cargo Loading and discharging to reduce the likelihood of over-stressing the hull structure

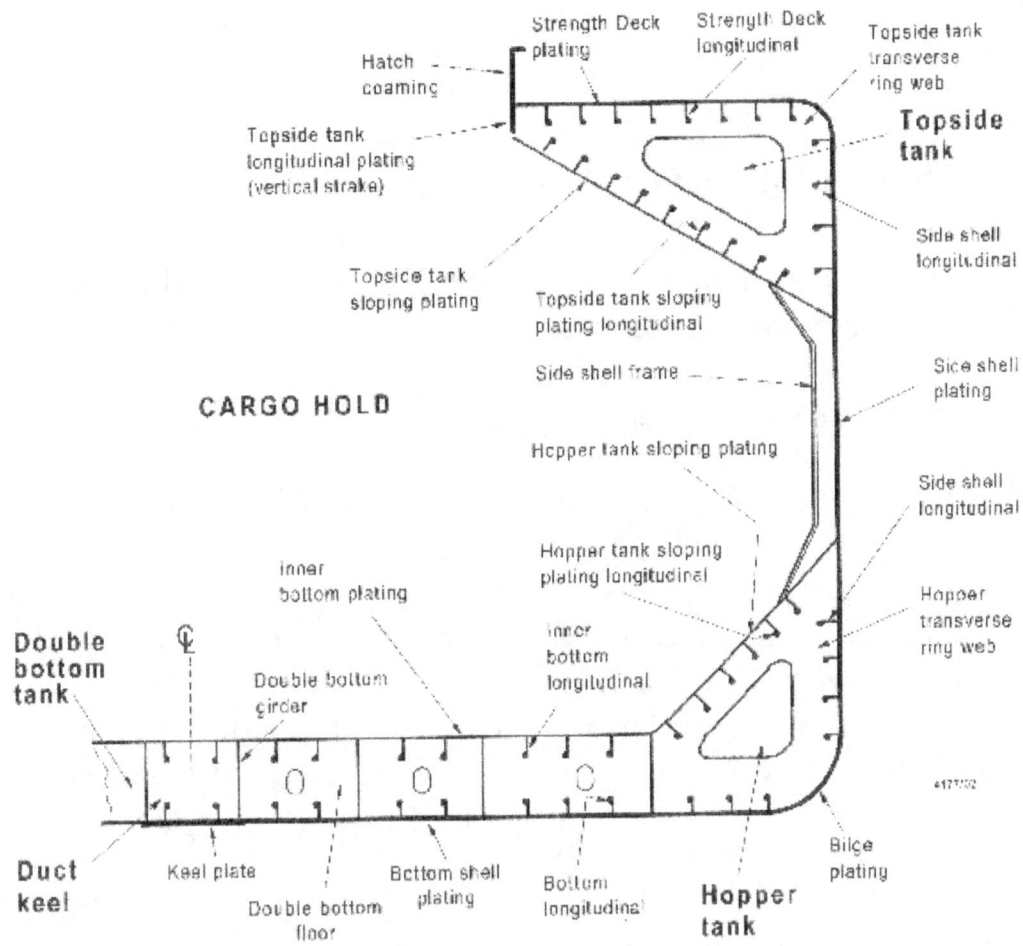

Illustration 2 Nomenclature of typical transverse Section in way of cargo (IACS -Bulk carrier: Guidance and information on Bulk Cargo Loading and discharging to reduce the likelihood of over-stressing the hull structure)

2.0 Preparation to take over the cargo

After each discharging the cargo hold has to beprepared for the next cargo. Out of a lot of experience the best procedure is:

a. Remove all residues inside the hatch

b. Remove all residues behind the piping systems and frames

c. Wash down the hatches with seawater. If high pressure seawater systems are available, using these systems is much better then only using the deck wash hoses, because the pressure on these high pressure cleaning systems is much higher compared to the normal pressure ofthe fire main or deck wash line. (6 bar). The advantage of a high pressure cleaning system is that dust and also harder particles can be washed out easily.

d. Clean all hatches with fresh water to remove the salt particles from the seawater.

e. Check that all frames, girders, beams and brackets in the hatches are clean and free of any residues from the previous cargo

f. Clean all bilges and check that the bilge suction is working. The bilges must be checked very intensively because the residues of the previous cargo will remain in the bilges and will, if not cleaned, create a bad odor which can spoil the next cargo. Check and test the bilge alarms and cover the bilge covers with burlap.

g. Make sure that the hatches will be fresh air ventilated to dry the hatches

h. The hatches must be free of any dust, salt, residues from the previous cargo. Prior loading, the hatches will be checked by a surveyor. If the ship is passing the survey check, the intended cargo can be loaded

After washing the hatches the bilge wells to be checks and cleaned. If necessary the bilges have to be covered for taking over the cargo. All girders, beams transfers frames, faces, brackets and stringers have to be checked if they are clean and no residues are found. Especially for grain cargoes the hatch must be free of any rust and any odour. The hatches should be good fresh air ventilated and free of any salt particulars. For this reason it is also important to clean up the bilges, because the residue from the previous cargo can spread a bad odour and can spoil the next cargo.

3.0 Important properties of the cargo

The properties of the bulk cargo to be loaded, should be well known by the ships command, in order to avoid damages to the cargo and to the vessel and last but not least to the crew.[1]

Bulk concentrates are very dangerous. The humidity in the cargo can result in a capsizing of the vessel, due to a high flow moisture point. If the humidity is higher or even with the flow moisture point, the cargo can start shifting. The indicator for the shifting can be either bad

---

[1] Steel, or steel plates are very dangerous for the crew if entering the hatches without ventilating.

weather or the rolling of the vessel in the sea, or high vibration of the vessel, caused for example by high and heavy head sea. For this reason it is a requirement to have reinforced longitudinal frames build in. These are extra high cost for building this reinforces longitudinal frames are girders. If the vessel will have already fixed longitudinal girder or wing tanks, the reinforcement is not necessary. The tremendous pressure on the bulkhead, if a concentrate is already shifted[2] should be not underestimated. Bulk carriers who have to carry concentrates as well should have a ship document, or certificate[3] that they can carry these cargoes. Another property is the self-ignition of concentrates, only if the will be shipped dry. Also oxidizing of concentrates is dangerous, because these cargoes will absorb the oxygen out of the cargo hold or space. If entering the hatch without BA, or air ventilating, the inspection can end up deadly.

For all these reasons it is important to know all properties of the cargo early in time and not after loading is completed.

To avoid accidents and damages to the ships and the crew, prior arriving in the loading port and prior loading the ships command must know the following figures.

- Stowage Factor of the cargo
- Density of the cargo
- Angle of repose
- Loading capacity of the loading device / hrs

---

[2] Will shift if the flow moisture point is over 10%

- Air draft on arrival
- Max. height of loading facility
- Depth at the berth
- Tides in Port
- Height of the Pier

These information can be found in the guide to port entry or you will receive the information from the charterer, together with the properties of the cargo.

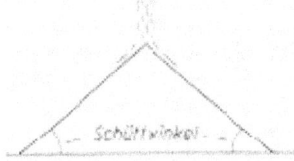

Illustration 3 Angle of repose - Source: P.Grunau

As you have now noticed, the properties of the cargo are from utmost importance.
For example coal: Oxidation and heat stow can be the reason for self-ignition of the coal. If loaded such coal, thermometers have to be placed into the thermometer pipes to monitor the temperature. If the temperature is above
40° C that indicates already a dangerous situation. All kinds of coal have the property to produce gases.

3.1 Angle of Repose

The angle of repose is the steepest angle that a stable slope can attain, therefore it describes the stability of the cargo.

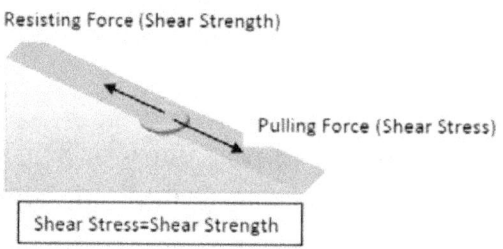

Illustration 4 Angle of repose - sheer stress and strength - Source: P.Grunau

3.2 Factors Affecting the Shear Strength that Determines the Angle of Repose

The factors which are affecting the shear strength are:

- ❖ Properties of particles – Size and Angularity

- ❖ Water content of the cargo

- ❖ Density of the cargo

a. *Size and Angularity of the Particles /Cargo*

Greater angularity of the particles will result in more intergranular friction and interlocking of the particles. That contributes to greater shear strength and, therefore, also to a greater angle of repose

b. Water Content

The water content affects the cohesiveness of the particles.

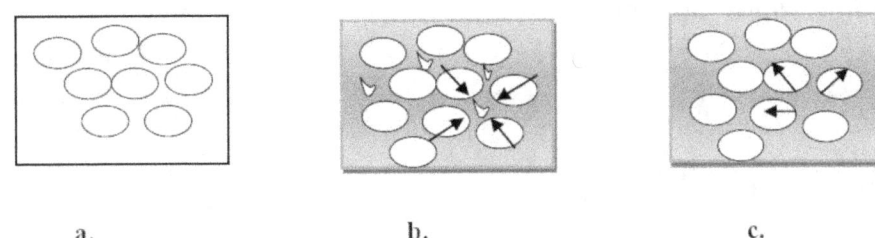

a.   b.   c.

Illustration 5 Cohesiveness of particles- Source P.Grunau

If water is added to particles, water coating grains tend to bend them together by its surface tension. This gives a greater internal cohesion and, therefore, more shear strength (Illust.b). But if water is added to completely saturate the pore space (Illust..c), the pore water will act as a lubricant between the grains, and the pore pressure will force the grains apart from each other, resulting in less shear strength and angle of repose. Illustration (a) presents the situation when no water is added. The pores are filled with air. The particles will interlock with each other and the angle of repose is smaller than when the cargo is wet.

3.3 Influence of the Angle of Repose to Ships' Stability

If the angle of repose of the cargo is taken for granted, there is a high risk of the cargo shifting by sliding.
Sliding occurs when the cohesive strength of the cargo is insufficient to withstand the effects of rolling. Shifting of the cargo always leads to a reduction of the stability of the vessel because the center of gravity is transversely shifted which causes the reduction of the stability.

Shifting of cargo can have numerous consequences:

☐ Structural damage

☐ Listing

☐ Capsizing of the ship

a. Structural damages
Dense cargo has a relative high mass to volume ratio. Small amount of shifted cargo can therefore have a large mass. Coupled with the momentum generated when a vessel is rolling, considerable forces can act upon the ships structure.
Repetitive shifting and sliding of cargo can result in plate flexing, increasing the risk of cracking and failure.

b. Listing

A vessel is listed when the cargo shifts to one side with subsequent vessel movement. The result is that there is an increase of draft on the listed side. The force returning the vessel from the angle of heel beyond the angle of list back to the same angle will be less than the force returning the vessel to upright if it has not been listed.
List will subject the vessel to greater angle of heel and this will reduce the ship's stability due to a drastic shift of the center of gravity in the transverse direction and also in the vertical direction.
When loading grain products, the shifting of the cargo can be reduced – if the hatches are not completely filled – by means of setting transverse or longitudinal center bulk head. These bulk heads will

reduce the space of the cargo hold and, therefore, also at the same time reduce the volumetric heeling moments of the compartment. The maximum angle of heel which should not be exceeded if loading grain is approximately 12°. Therefore, the risk of shifting cargo is also anchored in the cargo property itself. The property which is important for us is the non-cohesive cargo:

The characteristics of a non-cohesive bulk cargo are:

Non-cohesive bulk cargoes having an angle of repose less than or equal to 30°

The cargoes are flow-freely cargoes and should be carried according to the provisions applicable to the stowage of grain cargoes

Grain cargoes are:

Cargoes that need additional stability requirements (see also: "Stability Requirements for Grain Cargoes")

If necessary, the free surface must be secured

Trimming is essential

c. Capsizing

Capsizing will occur if the stability of the ship is in an unstable equilibrium. This might happen when loading bulk cargoes which tend to get liquefied. If they get liquefied, the vessel will get a sudden heel and most properly the vessel will also capsize because the stability is no longer sufficient to bring the ship back in upright. Therefore, special provisions applicable for the loading and stowage of these kinds of cargoes must be strictly observed.

4.0 Loading and discharging Procedures

To load and discharge large quantities of bulk cargo, an exact loading and discharging distribution plan have to be prepared. This plan includes the loading / discharging operation as well as the ballasting and de-ballasting operation. The sequences of ballasting and de-ballasting whilst loading different compartments have to be exactly followed and have to be discussed with the terminal manager prior loading. Due to the different kinds of Bulk Carrier (Multi - purpose - carrier[1], Bulk carrier with 7 and 9 hatches) also the loading and discharging plan is quite different. The 7 hatch and 9 hatch Bulk carrier are more sensitive for loading and discharging as the Multipurpose Bulk carrier

Illustration 6Multipurpose Bulk carrier, dischargingcoal in Hamburg(Source: P.Grunau)

---

[1] 3-5 hatches, but usually 5 hatches

But never mind what kind of vessel you will face, a loading and discharging plan have to be done.[1] To avoid a very high stress situation during loading and discharging this preplanning has to be done and are required. The terminal will not load or discharge the vessel if this plan was not discussed with the ships command.

Example : A Bulk carrier should load 45 000 tons of coal in the USA for North continent (Europe).The vessels have 7 hatches, 4 D.B tanks and 3 Wing-tanks . Each hatch has a volume with regards to the cargo of 8000 m^3. The density of the density of the cargo was measured with 1,26 m^3/mt - the S.F = 1,53 m^3/mt

According to this figures the vessel can load in each hatch a quantity of 6588,235 mt

([Volume / S.F] * density). 45000 tons should be loaded, equals 6 x 6588 mt and 1 x 5472 tons

The ship's command has now to clarify in which loading sequence, in accordance with the de-ballasting of the vessel, the vessel should be loaded. It has to put into account that the stress moments during loading and de-ballasting have to be screwed down to a minimum.

If the vessel will commence loading the ballast tanks a normally not empty and have to be de-ballasted during loading operation and ballasted during discharging operation. The Wing-tanks are usually de-ballasted by gravity, therefore the double bottom tanks have to be pumped out as soon as possible. For this reason the vessel must have good and acceptable trim by the stern. This must be part of the loading plan.

---

[1] During loading and discharging the vessel will be extremely stressed in all ships parts

Example of a loading plan

Cargo	Coal	S.F/Density	1,2	m³/mt	Total Volume		72.000	m³	POL	Richards Bay, S.A	Actual Ballast o/B	
Max. allowable Load-rate : 4200 mt		Loading Rate	mt/hr	3500,00	Total Cargo		60000,00	mt	POD	Rotterdam		13500,00
		Ballast Rate	t/hr	1200,00	Total Ballast		16800,00	m³	Max. Deballasting	14,00	Time	12,25

	TST No.3 P		TST No.2 P				TST No1 P		
Vol. m³	1800,00	Vol. m³	1500,00			Vol. m³	1500,00		
DBT 5P m³ 580,0	DBT 4 m³ 480,0	DBT 3P m³ 650,0	DBT 2	DBT2Pm³ 570,0	DBT 1Pm³ 420,0				
Volume 8650,00 m³	Volume 9650,00 m³	Volume 10200,00 m³	Volume 11800,00 m³	Volume 11720,00 m³	Volume 10420,00 m³	Volume 9560,00 m³			
Weight 7208,33 mt	Weight 8041,67 mt	Weight 8500,00 mt	Weight 9833,33 mt	Weight 9766,67 mt	Weight 8683,33 mt	Weight 7966,67 mt	1800 m²	VPTK	
Hatch No.7	Hatch No.6	Hatch No.5	Hatch No.4	Hatch No.3	Hatch No.2	Hartch No.1			
DBT 5S m³ 580,0	DBT 4 m³ 480,0	DBT 3 S m³ 650,0	DBT 2 m³ 570,0	DBT 1 Sm³ 420,0					
Vol. m³ 1800,00		Vol. m³ 1500,00		Vol. m³ 1500,00					
TST No.3 Stb		TST No2. Stb.		TST No1. Stb					

Pour	Hatch	Weight loaded [mt]	Deballast	[mt]	Time Load	Time Ballst	S.F [%]	B.M [%]	Draft Fw[m]	Aft[m]		
1	4	4000,00	DB 1P/S	1100	1,2	1,2	58	64	4,5	6,7		
2	1	3500,00	DB 1P/S	300	1,0	0,5	72	80	5,8	5,3		
			DB 2P/S	700	0,6	0,6						
3	6	5000,00	DB2 P/S	400		0,5						
			TST 1P/S	800	1,3	0,7	65	71	4,8	6,8		
			DB3P/S	600		0,6						
4	2	4000,00	DB3P/S	750	1,2	1,2	78	85	5,8	6		
			TST2P/S	1000								
5	7	5000,00	DB4P/S	960	1,3	1,3	76	77	5,3	7,9		
6	3	6000,00	DB5P/S	1160	1,5	1,5	75	72	6,2	7,5	Total Time Loading	15,90
7	5	7000,00	TST 3P/S	1000	1,6	1,6	73	79	6,4	8,6		
8	1	3966,00	Resting	200	1,1	1,1	73	76	7,6	8	Total Time Ballast	12,30

Pour	Hatch	Weight loaded [mt]	Deballast	[mt]	Time Load	Time Ballst	S.F [%]	B.M [%]	Draft Fw[m]	Aft[m]
9	6	3050,00	Resting	100	1,0	1,5	79	75	7,5	9,2
10	3	1760,00				0,5	75	79	8,5	9
11	7	2210,00			0,7		72	73	7,5	9,6
12	2	4690,00			1,4		68	74	9,4	9,2
13	5	1500,00			0,6		69	70	9,2	9,8
14	4	3880,00			1,5		65	71	9,6	9,6

Illustration 7 Cargo loading plan (Source: P.Grunau – Useful programs – Bulk cargo)

Illustration 8 Discharging grain via elevators (Source: P. Grunau)

The discharging will be conducted in the same way as loading. Also here a pre-discharging plan has to be prepared, because during discharging the same longitudinal stress moments occur than during loading.

To avoid a shifting of "fluid cargo" like grain or iron ore most bulk carriers are equipped with self-trimming hatches.

The cargo will be final measured via a draft survey and will be calculated by the Chief Officer and from charterers side, where a surveyor is acting for the charterer. On hand of a draft survey the cargo can be measured exactly, if all other weights are already known correctly.

In general bulk carriers are designed and approved to carry various kinds of cargo.

The most important point if loading cargo is to distribute the cargo that no extreme structural damages will occur. The most commonly adopted cargo distributions are:

- *Homogeneous hold loading condition*
- *Alternate hold loading condition*
- *Block hold loading condition*
- *Part hold loading condition*

4.1 Homogeneous Hold loading Condition (Fully loaded)

Homogeneous loading condition means the carriage of cargo, evenly distributed in all cargo holds. This kind of distribution is permitted for all bulk carriers and is usually adopted for carrying light, which means cargo with a low density, like coal or grain. But also heavy cargo, high density cargo, can be carried homogeneously sometimes. The picture below illustrates a homogeneous cargo distribution.

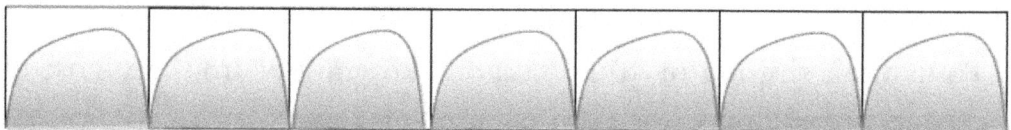

Illustration 9 Homogenous Loading (Source: P.Grunau: Cargo Handling and Stowage - A Guide for Loading, Handling, Stowage, Securing, and Transportation of Different Types of Cargoes, Except Liquid Cargoes and Gas

4.2 Alternate Hold loading condition (fully loaded)

Heavy cargo, high density cargo, will be often carried in alternate loading condition on bulk carrier. Alternate loading condition means, cargo distribution in odd numbered holds with remaining holds empty. This cargo distribution will raise the ship`s centre of gravity, which eases the ship's rolling motion. Important to know, if cargo will carrierd in this way of loading condition, the weight of cargo carried in each hold is about double that carried in a homogeneous condition. The local structure needs to be especially designed and reinforced, if carrying in an alternative loading condition. Ships which are not approved of heavy cargoes(High Density Cargoes) in alternate holds must not adopt this cargo load distribution.

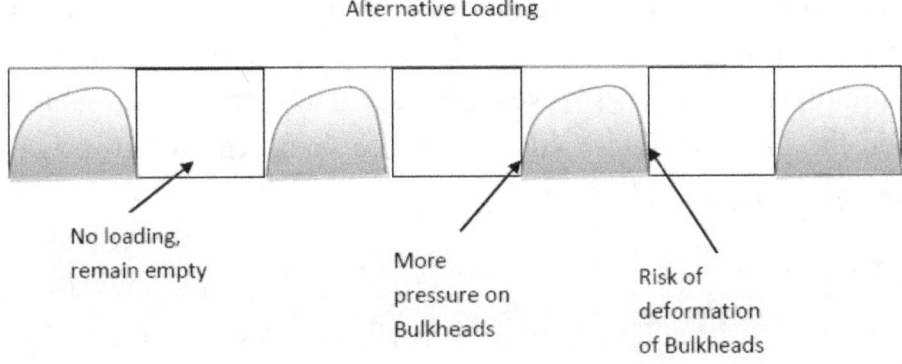

Illustration 10 Alternative Hold loading condition (Source: P.Grunau: Cargo Handling and Stowage - A Guide for Loading, Handling, Stowage, Securing, and Transportation of Different Types of Cargoes, Except Liquid Cargoes and Gas)

4.3 Block Hold loading and Part loaded condition

This refers to a stowage of cargo in a block of two or more adjoining cargo holds, with the cargo hold adjacent to the block of loaded cargo holds empty. This will be mostly adopted if the ship will be only partly loaded. These cargo distributions, part loading and block loading, are normally described in the loading manual. Important is that if a ship is loaded partly, over-stressing of the hull structure must be avoided, by considering the needs to be given to the amount loaded in each hatch and the anticipated sailing draught. The picture below shows a block hold loading.

The weight of cargo in each hold must be adequately supported by the buoyancy up thrust acting on the bottom shell. Here a reduction in draught causes a reduction in the buoyancy up thrust on the bottom shell to counteract the downward forces exerted by the cargo. If ships can carry cargo in block loading condition, the cross deck and double bottom structure needs to be reinforced, because block loading condition means higher stresses acting, in the transverse bulkhead between the block loaded holds. Therefore these kind of stowage can be adopted only, if:

- The loading distribution is described in the loading manual
- The ship is provided with a set of approved local criteria which define the maximum cargo weight limit as a function of ship's mean draught for each cargo hold and block of cargo holds

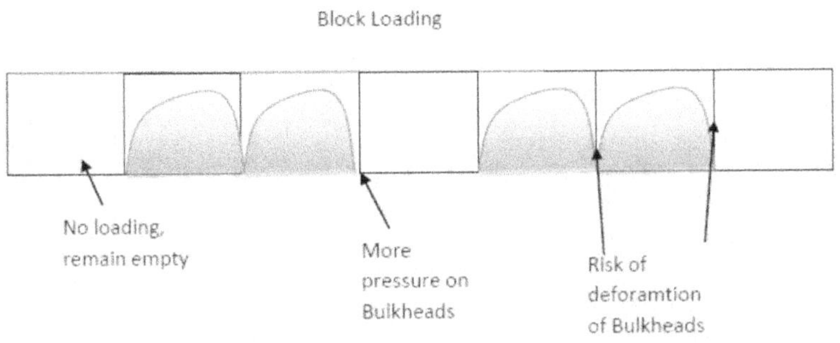

Block Loading

No loading, remain empty

More pressure on Bulkheads

Risk of deforamtion of Bulkheads

Illustration 11 Block loading condition (Source: P.Grunau: Cargo Handling and Stowage - A Guide for Loading, Handling, Stowage, Securing, and Transportation of Different Types of Cargoes, Except Liquid Cargoes and Gas)

Never mind which loading condition - distribution the ship command will use, they always have to pay attention to the longitudinal stress condition of the vessel, during loading, discharging and the voyage, besides the stability condition. The SWSF and SWBM (Still water shear force and Still water bending moment. During loading and discharging there are different forces acting along the ship's length. These are local differences in the vertical forces of buoyancy and the ship's weight. If these forces are unbalanced, these vertical forces will cause the ship hull girder to shear and bend. For this reason the SWSF and SWBM must be correctly calculated during loading and discharging, because as already seen, along with loading and discharging the ballasting and de-ballasting to be done, which has mainly a strong influence on these forces.

See illustration below for shear forces and bending moments and the limits which should not be exceeded as per example.

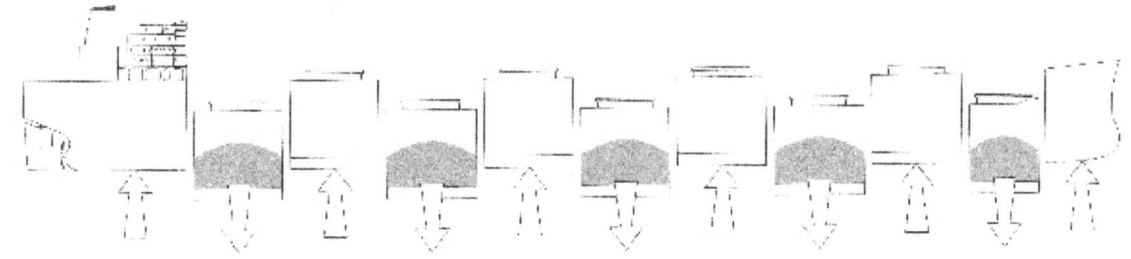

Illustration 12 Shearing action of the hull girder in still water

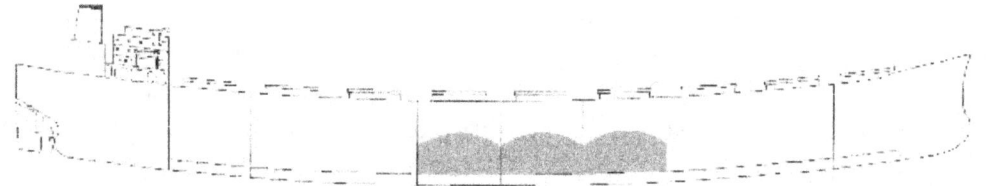

Illustration 13 Bending of the Hull girder - sagging

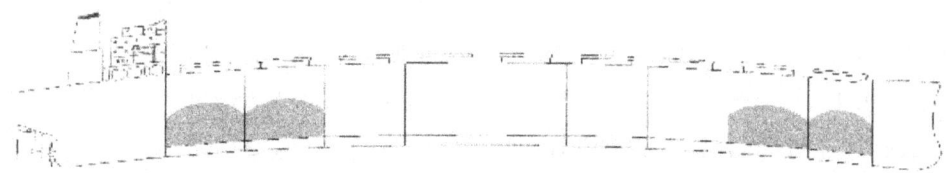

Illustration 14 Bending of the Hull girder - hogging

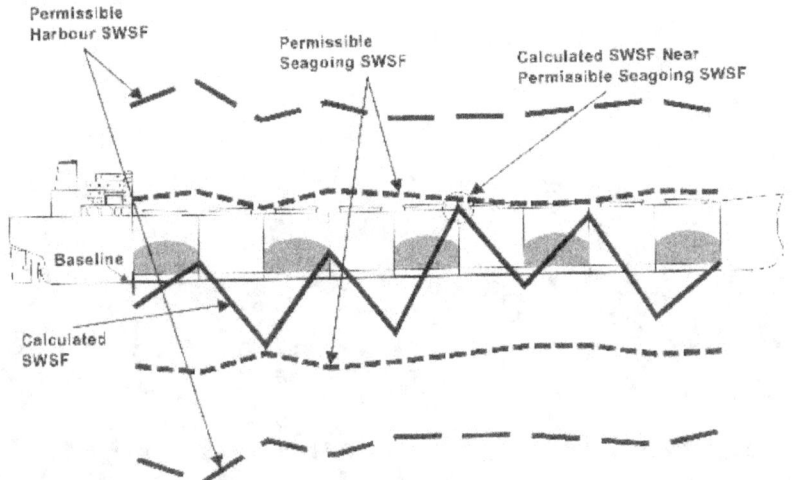

Illustration 15 Permissible Stress - still water stress moments
(IACS -Bulk carrier: Guidance and information on Bulk Cargo Loading and discharging to reduce the likelihood of over-stressing the hull structure)

During loading of bulk cargo it is advisable to load in accordance with the permissible limits of the Still Water Stress condition.(SWSF and SWBM). The harbour stress moments are always higher in the limits and are not representing the seagoing condition. If using a load master – computer program- the loading should be done in seagoing condition.

4.4 Load planning using a Load Master

On hand of an example the single steps, which were already explained, using the load plan template, will be presented. The pours will presented step by step. The vessel is a seven hatch bulk carrier.

Cargo: Coal   **Density: 1, 20**

Loading Rate: 6000 mt / hrs.

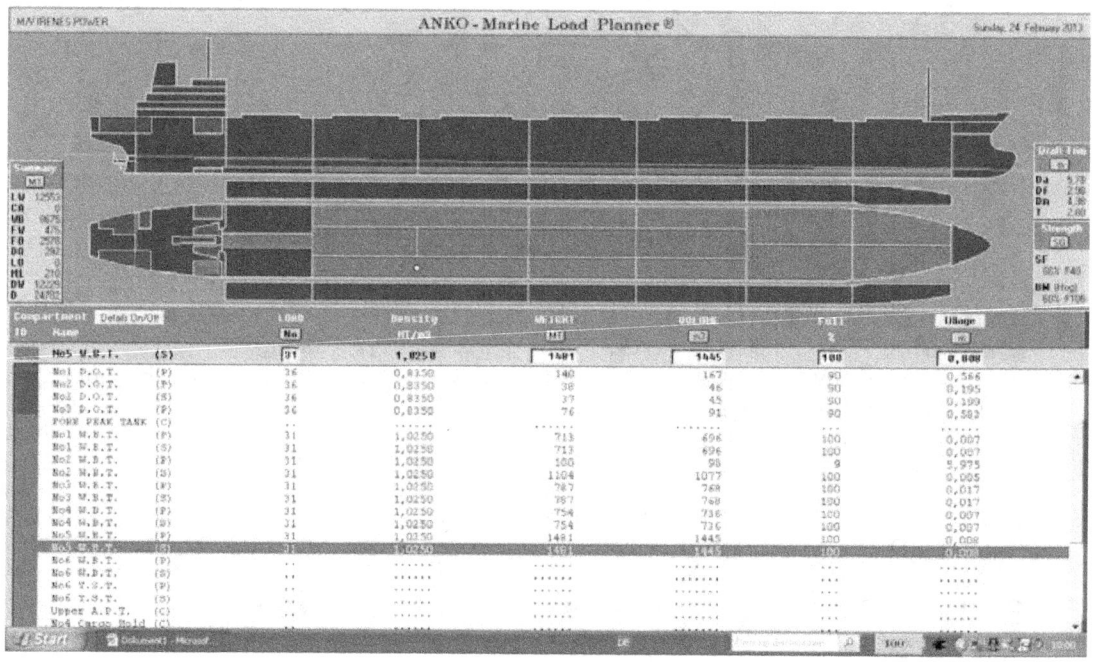

Pour No 1: Loading 1 hr. = 6000 mt Hatch No 4- De-ballast: BWTK 1 P/S

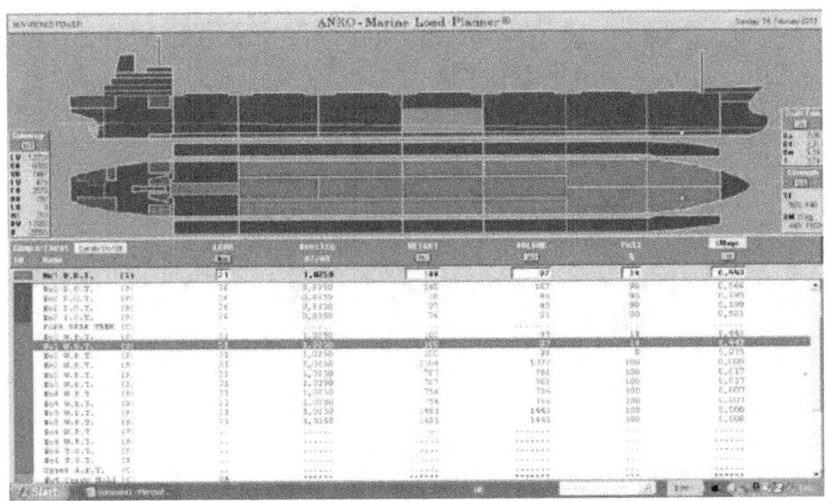

Pour No 2: Loading Hatch No 1 = 6000 mt -De-ballasting WBTK 2 – resting WBTK 1

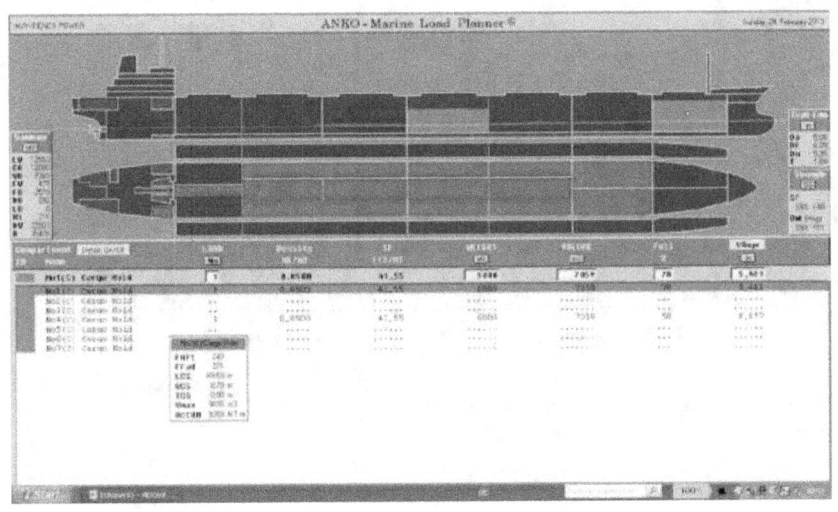

Pour No 3: Loading Hatch No 7= 5500 mt - De-ballasting WBTK 2

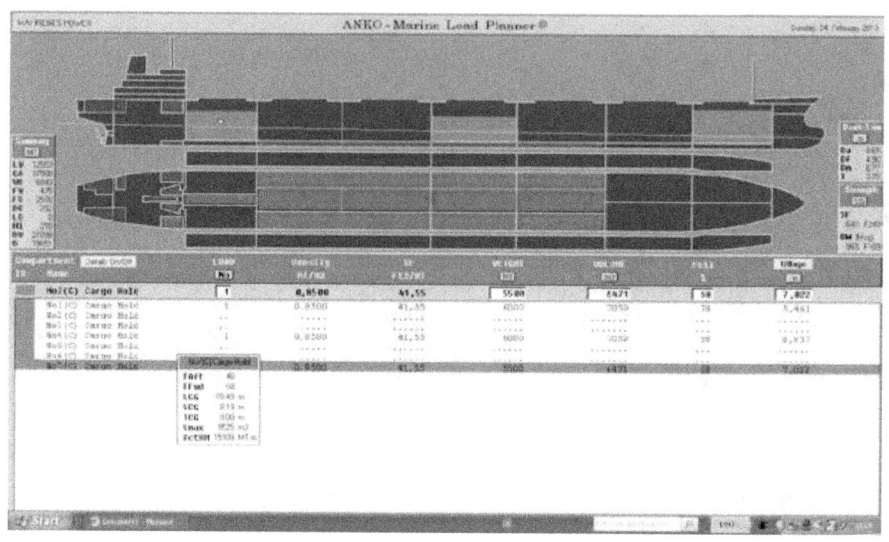

Pour No 4: Loading Hatch No 3 = 6000 mt - De-ballasting No 3

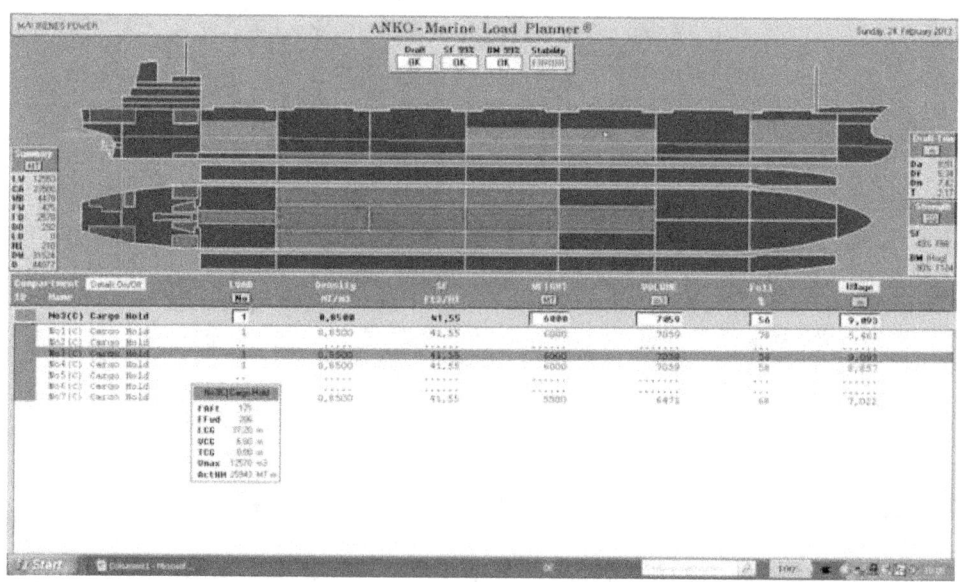

Pour No 5: Loading No 6 = 8000 mt - De-ballasting No 4

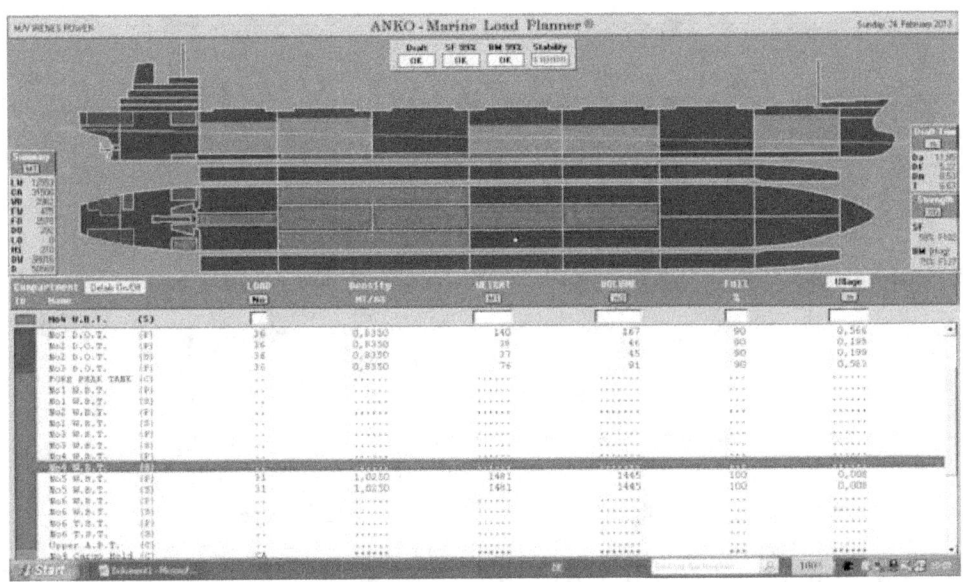

Pour No 6: Loading Hatch No 5 = 5400 -Start de-ballasting WBTK 5 P/S (50%)

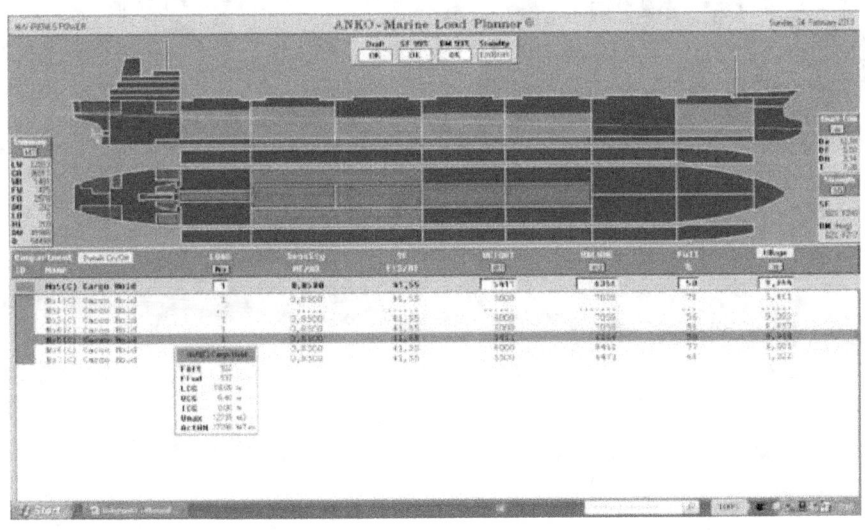

Pour No 7: Hatch No2 Full - Emptying WBTK 5 P/S

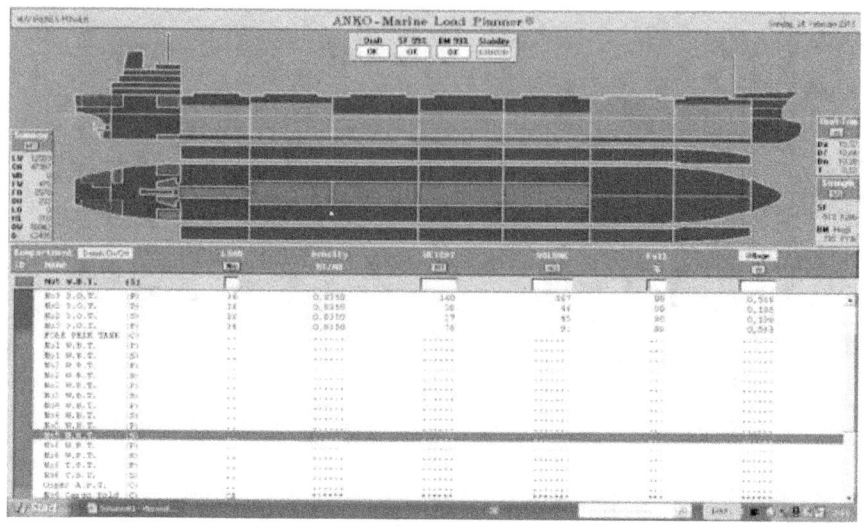

Pour No 8: Hatch No 7 Full -Stripping WBTK's

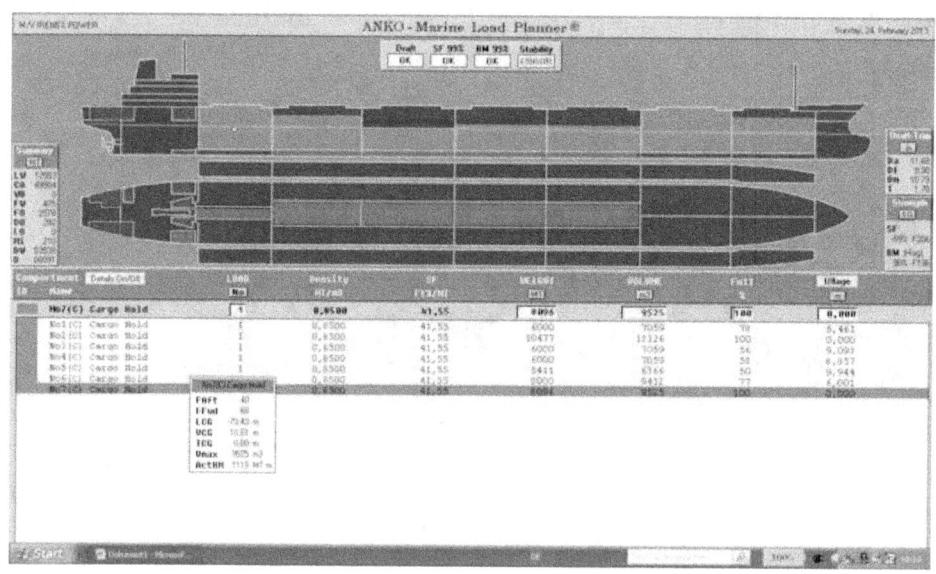

Pour No 9: Hatch No 3 Full

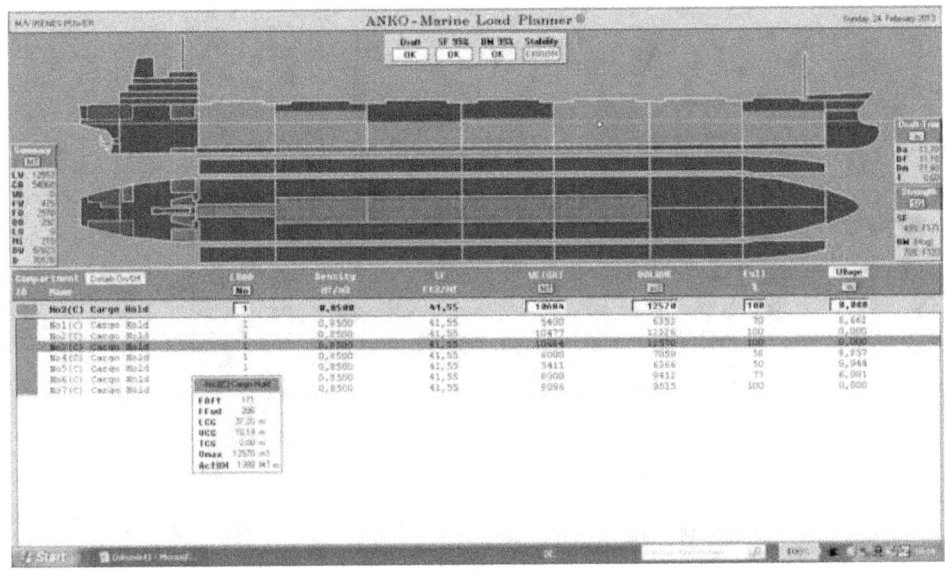

Pour No 10: Hatch No 5 Full

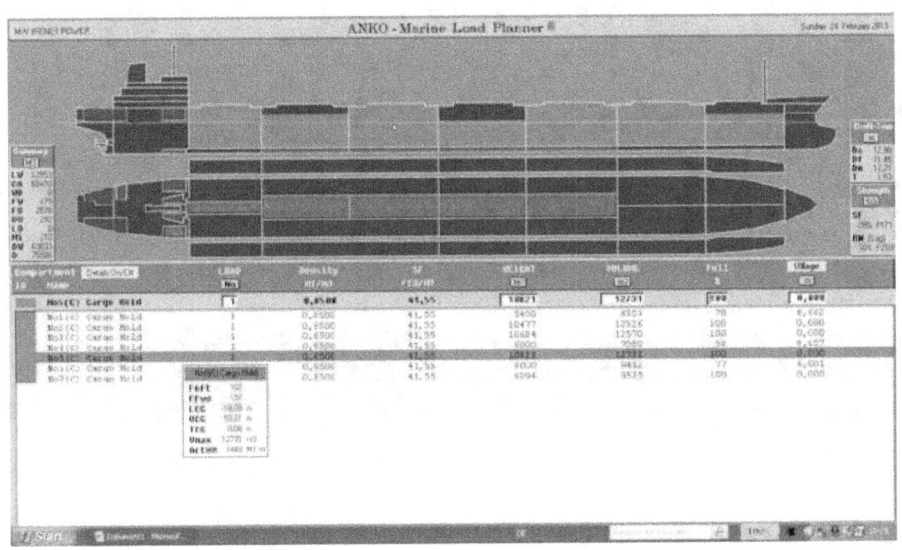

Pour No 11: Hatch No 1 Full

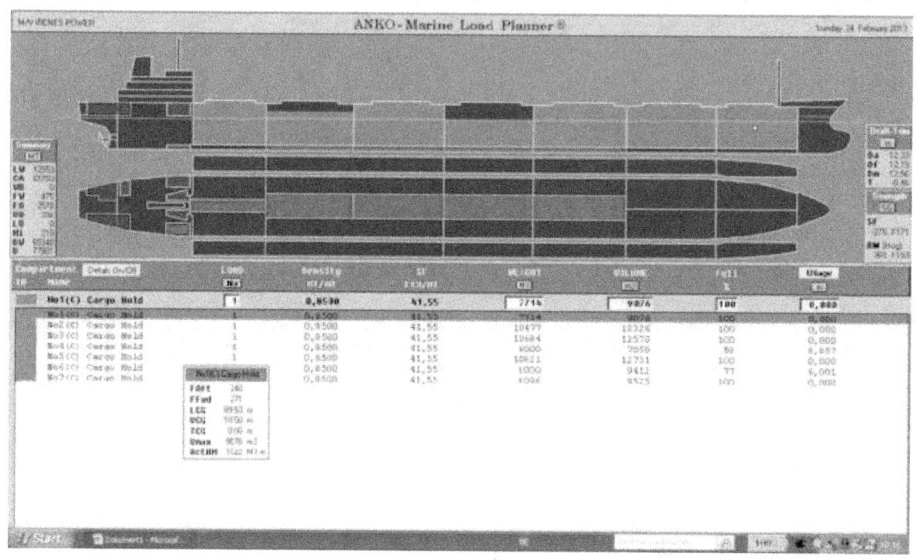

Pour No 12: Hatch No 6 = Full

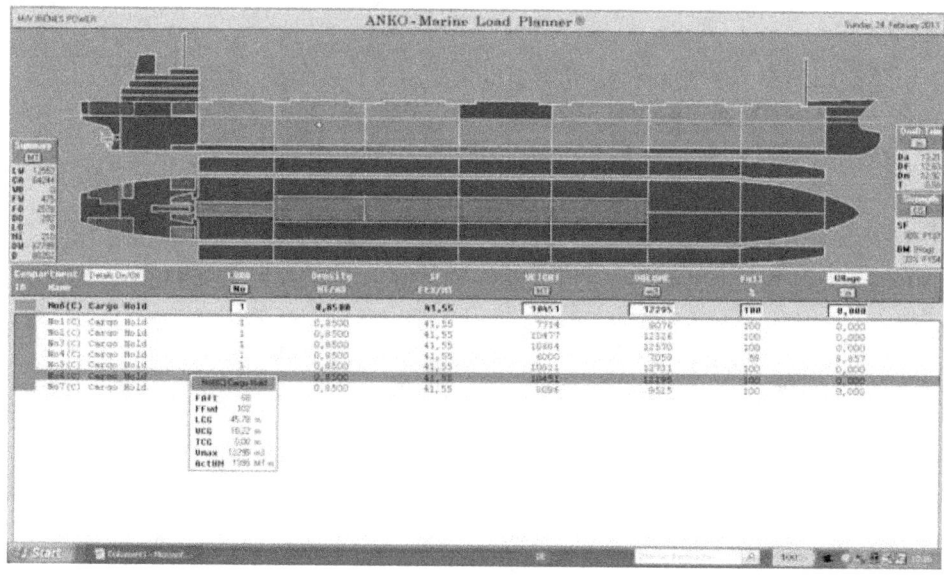

Pour No 13: Rest Cargo in Hatch No 4 (Full) to bring Ship on draft

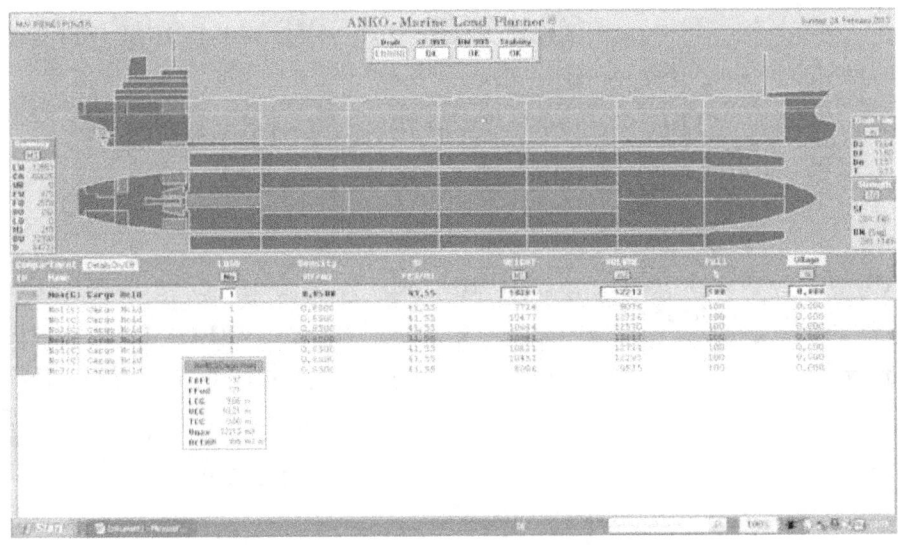

Illustration 16 Load planning by means of a software (All illustration are screenshots of the Program: ANKO Marine Load Planner for MV Irene Power

Total Cargo Loaded: 68625 mt

Loading Time: 11,4 hrs. + 0,6 hrs. for trimming = 12 hrs.

4.5 On board loading Guidance

It is a requirement (International load line convention) that the master of a ship must be supplied with sufficient information in approved from to enable the master to load and ballast the vessel, that no unacceptable stresses in the ship's structure occur. Therefore a ship's approved loading manual is essential. The documentation must describe loading, discharging and stowage operation. It should contain the following

- Loading conditions on which the design of the ship has been based, which includes the permissible limits of the longitudinal stresses.

- The result of SWSF and SWBM calculations for each loading condition
- The allowable local loading of the structure
- Operational limits

Another import part for loading and unloading of bulk cargo on bulk carriers are the loading instruments. These are individual shipboard calculation tools, which will assist the ship command in loading and discharging of the vessel. They are never a substitute to the loading manual. The loading instruments must enable the ship command to:

- Plan and control the cargo and ballast operation
- Rapidly calculate the SWSF and SWBM for any loading condition
- Identifying the imposed structural limits which are not to be exceeded

4.6 Ratio between Loading Rate and De-Ballast Rate

Bulk carriers cannot be loaded at any loading rate. The loading rate needs to correspond with the ship, the cargo to be loaded and the allowable strength condition of this vessel. Further, the loading rate must be in accordance with the de-ballasting rate of the vessel, not to overdue the allowable strength condition of the vessel. Therefore, there is a ratio between loading rate and the maximum de-ballasting rate.

Comparing a Bulk Carrier with a Reefer Vessel or General Cargo Vessel, the ratio of the minimum ballast water condition and the maximum deadweight at this minimum ballast water is equal to 1:7.

The ratio at maximum ballast condition to maximum deadweight is 1:4. The ship command must check what the maximum loading rate compared to the maximum de-ballast rate is (let always both ballast pumps running) to guarantee a loading of the vessel without interruption and not exceeding the allowable strength condition of the vessel during loading.

In the past it was empirically proofed that the maximum ratio: Loading rate – de-ballasting rate, in accordance to the strength and actual ballast condition of the ship should not be more than 1:3, 5. At a ratio which is above 1:3,5 , an uninterruptible loading is no longer possible. The amount and weight of the minimum ballast condition on board and the fast increase of cargo weight will exceed the maximum allowable strength condition of the ship.

The ratio loading rate to de-ballasting rate plays an increasingly important role. The loading devices ashore can keep loading rates of 14000 – 16000 tons/hr., some can even make more.

For example: If a 150.000 tons Bulk Carrier will be loaded with a loading rate of 16000 tons/hr., the ship will complete the loading operation after ~ 10 hrs. Let's assume the vessel has two ballast pumps with together 3000 m³/hr. In total, the complete de-ballasting operation without resting the tanks will be 6,8 hrs. If we calculate another 1,5 hours for the resting of all tanks, the complete de-ballasting time will be 8,3 hours. Normally we could say the de-ballasting operation will be completed 1,7 hours prior completion of the cargo operation, so a loading without interruption should be guaranteed. But this is not the case if we are looking at the ratio of loading rate to de-ballasting rate because this rate is 1:5,3. This will by far exceed the allowable ratio of 1:3,5 where an uninterruptible loading is still possible in accordance to the allowable stress condition of the vessel.

What is now possible?

1. Theoretically, the ship can be equipped with ballast pumps which will do 5000 m³/hr. (with both or all ballast pumps). Then the maximum loading rate can be 17500 tons/hr. The ratio will still be 1:3,5. To be on the save side, the ratio should be below 1:3,5.

The loading rate must be in accordance with the de-ballasting rate and should not exceed the ratio 1:3,5 at a fixed maximum de-ballast rate. In our example, the de-ballast rate was 3000m³/hr. To keep the ratio 1:3,5 , a maximum acceptable loading rate is 10500 tons/hr. To be below the ratio of 1:3,5 , the loading rate, which can be agreed, is 10000 tons/hr. because at a maximum de-ballast rate of 3000 m³/hr. the ratio equals 1:3,3. As we already have seen, the ratio is based on the minimum ballast water for such a loading. The loading rate should also not exceed the maximum allowable strength condition. Even if we could say it is granted with this ratio, a re-check must be done (pre-calculation of the single pours and the strength condition occurring at these pours). If this calculation is satisfactory, the loading rate can be accepted.

Reminder:
Always be aware that elder bulk carriers are getting weaker in their structural parts, resulting in a poorer strength condition. A continuous overstressing of the structural parts will also weaken and reduce the strength condition of a bulk carrier.

4.7 Loading program and de-ballasting program for loading of Bulk Carrier

1. The first pour should, if possible, be loaded into a midship or after hold to provide or maintain a reasonable trim by the stern for de-ballast and stripping purposes.

2. If the air draft is restricted it will be necessary to make the first pour into a hold which causes some increase in forward draft to ensure that the loading spout can continue to clear the heath coamings of the forward holds.

3. If the air draft is restricted the effect of a rising tide must be considered and de-ballasting cannot continue when the clearance is small.

4. Successive pours should alternate between forward and after holds to maintain a reasonable trim by the stern.

5. The end holds (i.e., the foremost and aftermost holds) have the biggest effect upon trim. Where possible they should receive the last pours of the first pass, and the first pours of the second pass, because the resulting large changes in the trim and maximum draft are likely to be least inconvenient at that point.

6. The ballast which is likely to present most problems should be discharged first, the normal sequence commencing with ballast holds, continuing with double bottom tanks and wing tanks and concluding with peak tanks.

7. Ballast should normally be discharged from a position close to the one where the cargo is being loaded at that time. For example, No.3 double bottom should be discharged whilst No. 3 hold is being filled, if No.3 double bottom is below No.3 hold. De-Ballasting can be also conducted from fwd to aft if the vessel will be not overstresses.

8. The time required for a de-ballasting step should be matched with the time required for a loading pour. A pour of 3,000 tons at a loading rate of 1,500tons/hour will take two hours. This should be programmed with a de-ballasting step which will take less than

two hours, so as to reduce the likelihood that the de-ballasting will overrun, and become out of step with the loading.

9. The ballasting should be programmed to be completed several hours, at least, before completion of loading, and at a time when the vessel still has a stern trim, to assist the de-ballasting and stripping.

10. On many bulk carriers trim can be quickly and conveniently changed by pumping ballast directly from forepeak to after peak, or vice versa.

11. Rules imposed by the Classification Society and quoted in the loading manual may restrict the sequence of loading: they must be strictly observed. For example the manual may state that no hold can be completely filled until the mean draft is at least two thirds of the intended sailing draft.

12. In exposed berths the ship should be maintained at adraft and trim at which she can put to sea at short notice if required. This precaution is particularly recommended in areas where ports must be evacuated on the approach of a tropical storm.

4.7.1 Loading with two loaders at the same time

When two loaders are available the ship is divided into two, and each loader works its own end. The loading program devised for a single loader is normally suitable for two-loader loading, provided that the rate of de-ballasting is sufficiently high and that the original de-ballasting program can be followed in step with the loading. As with loadings with a single loader, stern trim should be maintained, but extremes of trim that may cause clearance difficulties should be avoided. See below steps contains a typical loading for the MV *Iron*

Somersby (Bulk carrier practice -A practical guide) for one-loader operation and the same loading adapted for two loaders.

It must be remembered that a second loader does not necessarily mean a doubling of the loading rate that two loaders are seldom available for the entire loading and that variations in the pouring rate are likely. Extra vigilance is required when different grades are being loaded.

When loading with two loaders both must plumb the center line to avoid twisting the ship's hull. This is most important as a ship which is kept upright by loading to starboard of the centerline in No.3 hold, say, and to port of the centerline in No. 7 hold will be subjected to cargo torque, or twisting of the ship, which may cause serious damage to her structure. If troubles with ballast occur the first response should be to stop one loader.

Because of their size, loaders usually require at least one hatch between them. This should cause no problem to the ship as it is generally undesirable to load adjacent hatches. When it is not certain whether one or two loaders will be used it is prudent to plan for two, adopting a plan which is also suitable for one loader. At some grain terminals up to five spouts may be used to load. Loading a low-density cargo in so many positions simultaneously will eliminate trim and longitudinal stress problems provided that a sensible distribution of cargo loaded is adopted.

IRON SOMERSBY (1976) LOADING PROCEDURE

Run	Hold	Tonnes	Ballast
			After draft survey dump all wing tanks
1	9	8,000	Pump No. 5 hopper (Pt & Stbd)
2	5	10,000	Pump No. 3 hopper (Pt & Stbd)
3	3	10,000	Pump No. 2 and No. 1 hoppers (Pt & Stbd)
4	7	12,000	Strip as required.
5	1	12,000	Ballast out (8 hours) Draft check
6	9	7,000	(If ballast is slow, pouring may continue to end of Run 9, when loading
7	5	8,000	should cease until stripping is completed).
8	3	8,000	
9	7	6,000	
10	1	5,500	
11	9	7,000	
12	5	4,700	Draft check
13	3	2,500	
14	7	3,000	
15	3	1,000	
16	7	1,300	
		106,000	

One loader at 6,000 TPH takes around 18 hours.

Two-loader operation

Loader No. 1				Loader No. 2			
Run	Hold	Tonnes	Elapsed Time	Elapsed Time	Run	Hold	Tonnes
1	9	8,000	0120	0140	2	5	10,000
4	7	12,000	0320	0320	3	3	10,000
6	9	7,000	0430	0520	5	1	12,000
7	5	8,000	0550	0640	8	3	8,000
9	7	6,000	0650	0740	10	1	5,500
11	9	7,000	0800	0820	12	5	4,700
14	7	3,000	0830	0845	13	3	2,500

Identical ballast sequence is followed.

On the following page is a diagrammatic explanation of the two-loader operation, in an hour by hour sequence

(Courtesy BHP)

Illustration 17 Loading sequence with one and two loaders(Reference: Bulk Carrier Practice - A practical Guide - Capt. J.Isbester - The Nautical Institute)

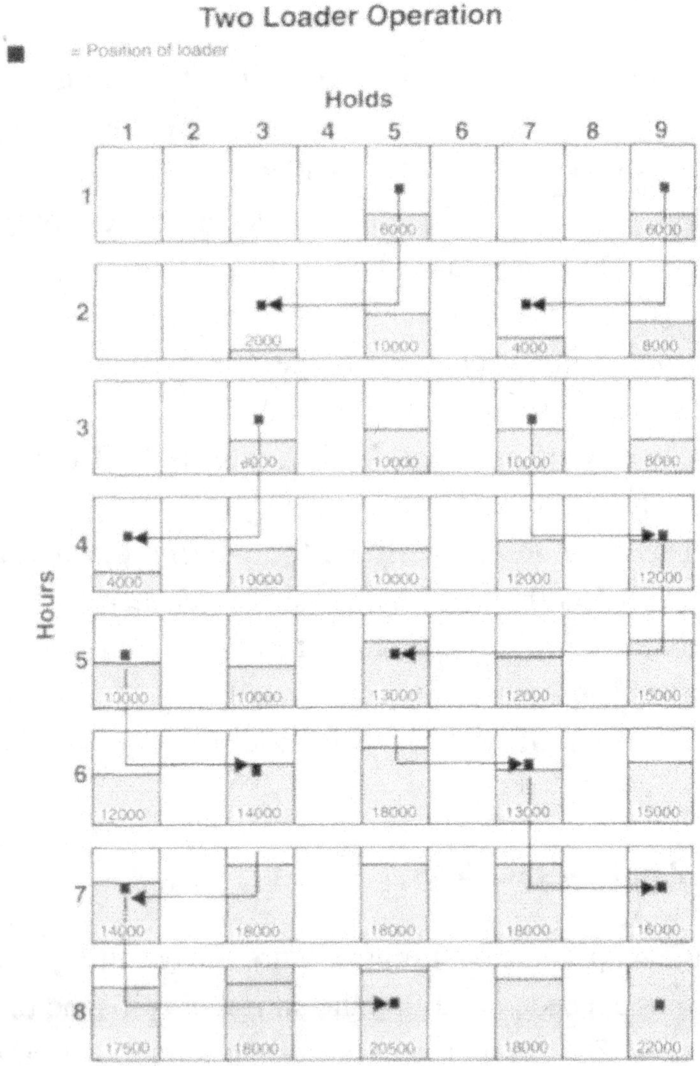

Illustration 18 Loading Procedure if two loaders will be used(Reference: Bulk Carrier Practice - A practical Guide - Capt. J.Isbester - The Nautical Institute)

5.0 Maintenance during the voyage

In general the cargo has to be monitored daily. It depends of the kind of cargo, but air ventilation and temperature checks are very important. (See also coal - Temperature above 40°C ......)
A hold control should be only done if all risks are under control and if not otherwise possible, a BA has to be present.
For some cargoes also the CO^2 level has to be monitored.
The ship's command must realize that the relative humidity will not increase, see also concentrated cargo in bulk. The cargo bilges should be stripped every day. (Important for iron ore and all other cargoes which were already wet loaded.)
Another point has to be observed all the time during the voyage to the discharging port, this is the stability of the vessel. Especially if I had loaded concentrates or grain. The risk that cargo will shift is still there, SO the stability of the vessel has to be in all respects in accordance with the requirements of the classification society. (All over the voyage).

5.1 Limiting Points of the Voyage

Prior loading the ship command must ensure that limiting points of a voyage will be not exceeded and that the ship will fulfil the mandatory legal requirements. The limits may restrict the amount of cargo to be loaded at the port of loading. What are the limits which might restrict the cargo the ship can carry?

- Maximum permitted draught at each stage of the voyage. The restriction depending on the geographical zone the ship is passing, the time of the year and the corresponding load line zone.

- Intermediate port during the voyage, such as bunker port, canals or other waterways, where the draught is restricted
- Draught restriction in the port of loading and discharging
- Tidal condition in the port of loading and discharging
- Weather condition during the voyage - depends on the season - like Tropical revolving storm season in the North Atlantic, Pacific and or Indian Ocean.

The Ship command must consider all these limiting points and the total cargo to be loaded must be based on the before calculated limits.

Example:

5 Hatch Bulk Carrier with Gears

The ship has to load a full consignment of Coal = 40000 mt - in Norfolk for Rotterdam - North Continent.
Time of loading: 27.March 2015
The pre-calculated displacement = 56200 mt
Lightship weight: 11079.5 mt
Deductible weights: 5120.5 mt
Mean Draft: 11.30 m
The consignment to be loaded = summer draught
Max. Voyage speed= 13.5 knots
Consumption: HFO: 30 mt / day
DO: 1 mt/day
Summer draught: 11.35 m
Winter draught: 9.95 m
TPC for the planned displacement for 11.30 m = 56.2 t/cm

Total distance Norfolk to Rotterdam - Europort = 3509 nm

Limiting Points of the Voyage

Norfolk Port: No draught restriction for the vessel during loading
Rotterdam -Europort: No draught restriction on arrival

The voyage:
The intended route is:

Illustration 19: Intended Route from Norfolk to Rotterdam

From Norfolk and after passing the approaches of Norfolk the Master intends a NE'ly course until WP3. From WP 3 to WP 4 Great Circle sailing where the vertex of the great circle = 50°02' N and 20° 01,5' W.
From WP 4 following the English Canal - Dover Strait until Rotterdam
This is navigational wise the shortest distance.

But if preparing the limiting point of the voyage the ship command must take into account the legal requirements for this voyage as well. Do I comply with the load line zones if I would follow the intended route?

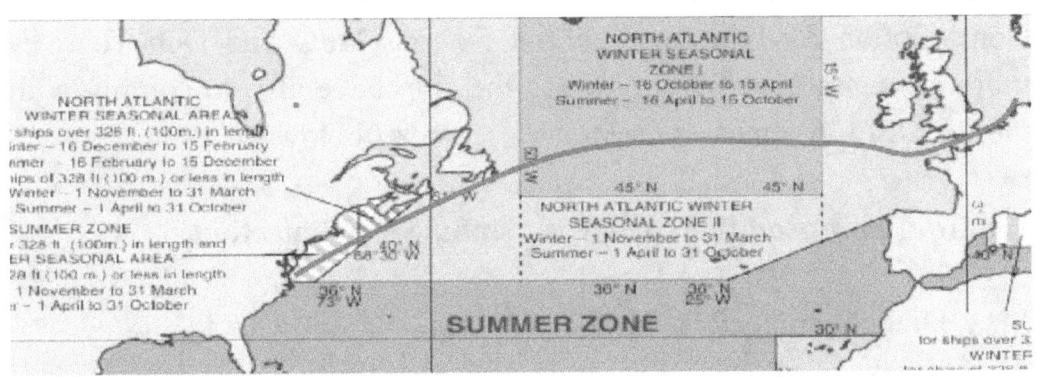

Illustration 20: Extract form Load line chart for the intended voyage (Reference: forum.shipsplotting.com)

Transferring the intended route into the load line chart the vessel will transit 3 load line zones:

Summer zone
Winter zone - Seasonal Zone I and
Winter zone Seasonal II
It is obviously that as soon as I reach WP 3, to start the GC, the vessel is already in the winter seasonal zone II - Winter zone from 1.November until 31.March
Further the vessel will transit the winter seasonal zone I - Winter zone from 18.october until 15. April
Therefore about one day after leaving Norfolk, which is in the summer zone - 01April to 31.October, the ship have to pass the winter zone.

According to the load line rules the ship have comply with the load line regulation as soon entering the respective load line zone (based on the Mean draught).

Draught on departure Norfolk is 11.30 m. On arrival in the winter zone the draught should be 9.95 m - winter load line for the ship.

Consumption for one day is about 56 tons HFO and DO. There is more consumption of DO because the ship have also to comply with the ECA/SECA zone requirements - see chart below

Limiting Points on the voyage:		max:Ship limitation
Max. Draught at Norfolk Berth max. 11.30 m	-	17.5 m
Norfolk Channel max. 11.30 m	-	19.0 m
Entering Summer Zone max. 11.30 m		11.30 m
Entering Winter Zone max. 11.29 m	-	9.95 m
Europort max. 11.23 m	-	22.25 m

Limitations are not fulfilled

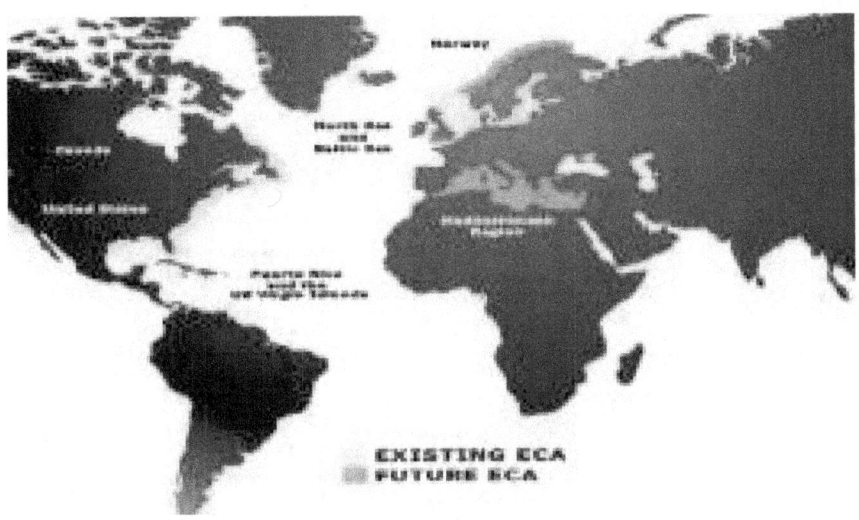

Illustration 21 : ECA Zones of the world (Reference: http://marineurea.com/marpol-nox-regulation)

If we consume 56 mt in total until reaching the winter zone, the rising of the ship is 1 cm, because the TPC is 56t/cm, equals to a mean draught of 11.29 - not fulfilling the requirement for the winter load line.

The displacement for winter load line is 49970.0mt. Deducting the light ship and the other deductibles the maximum weight to be loaded in Norfolk is 33770 mt. In addition a can load 56 mt more which I will consume until I reach the winter load line, therefore the total cargo to be loaded in Norfolk is 33826.0 mt.

An alternative route would be:

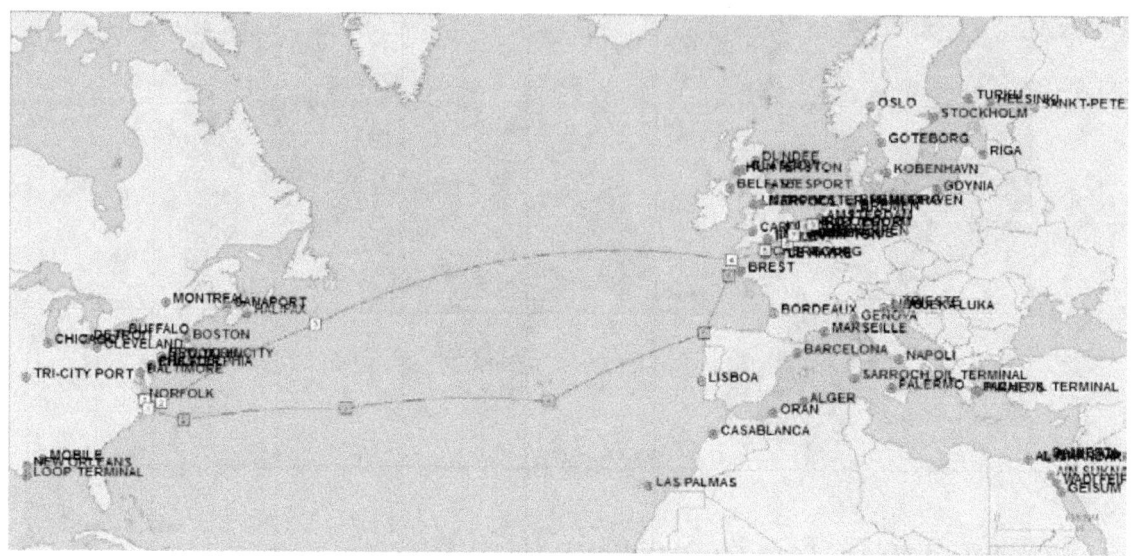

Illustration 22 Alternative Route - Southern Route

After departure Norfolk, a southern GC will be used. The advantage is I will navigate in the summer zone until arrival at the Gulf of Biscay, which is winter zone. Therefore I can take more cargo.

Total distance from Norfolk to Rotterdam is 3979.1 nm. Compared to the origin intended route the difference is 470.1 nm. At an average speed of 13.5 knots we will sail 35 hrs. longer than intended.

If the charterer would agree, the ship can load more cargo, means the consumption from Norfolk until Cape Finistere can be loaded in addition. From Norfolk to Cape Finistere the total distance is 3282 nm. At a speed of 13.5 knots and a consumption of 30 mt HFO and 1 ton Do / day this will result in:

$$t = \frac{d}{v} = \frac{3282 nm}{13.5 knots} = 243.11 hrs = 10.13 days$$

$$Total\,consumption = Cons.HFO + DO * t = 30mt + 1mt * 10.13 days = 314.03 mt$$

Also here the limiting points of the voyage are the load line zones. The cargo to be loaded in Norfolk must be accordingly, therefore the ship can load in Norfolk until winter load line+314 mt = 34084 mt. Until cape Finistere the vessel will stay in the summer zone - with our draft on departure of 9.95 m (winter load line) +~6 cm for the additional cargo which can be loaded for the consumption we will have a final mean draft of 10.01 m.

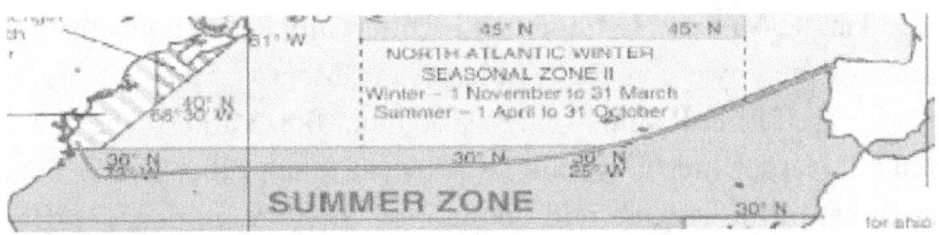

Illustration 23 Alternative Route transferred into Load Line Chart (Reference: forum.shipsplotting.com)

Limiting Points of the voyage:		**max: Ship actual**
Max. Draught at Norfolk Berth max. 10.01 m	-	17.5 m
Norfolk Channel max. 10.01 m	-	19.0 m
Entering Summer Zone max. 10.00 m		11.30 m
Entering Winter Zone max. 9.95 m	-	**9.95 m**
Europort max. 9.94 m	-	22.25 m

All required limitations are fulfilled.

5.2 Voyage Planning

Voyage planning is, regardless of the type of cargo or ship, a mandatory requirement. The IMO resolution 893(21) is requiring a detailed voyage plan for the intended voyage, and also the different rules, like Hague-Visby, Hamburg and Rotterdam Rule requiring the preparation of the voyage as part of the seaworthiness of the ship.

Bulk Carrier are full hull ships - large Cb, about 0,8- values. The ships tend to have a larger inertia (moment of inertia) than other ships. This must be known if the voyage will be planned. Areas with dense traffic - fishing boats, should be avoided if possible and a clear distance should be given.

The stress situation of Bulk Carriers are different compared to other ships during the voyage. High dynamic stress will result in higher bending moments and shear forces during the voyage, especially if navigating in severe weather condition. The voyage plan must consider these facts. This is important if trading in winter times, North Atlantic, North and Baltic Sea or in seasons where tropical storms are present.

The master must ensure that loss of cargo or damage of cargo must be avoided. (See also Hague-Visby Rule, Hamburg and Rotterdam Rule).A not well planned voyage can be considered as a ship management error, where carrier and ship are not liable for any claims rising in loss or damage of cargo.

For example: A full consignment of Coal was loaded in Africa with destination North Continent. The voyage is from warm to cold. The voyage must be planned that the ship will not pass upwelling cold water, which will lead to ships sweat and will affect the cargo, which might damage the whole or partly cargo.

Severe weather condition during the voyage might affect the cargo due to sea sloshing - cargo might be affect by sea water.

For cargoes like steel cargoes or grain cargoes this is important. Steel cargoes should not exceed a certain amount of humidity, therefore ventilation of the cargo should be carefully considered. This is part of the voyage plan, too.

The voyage plan is a risk analysis of the intended voyage. This analysis will contain:

- Terrestrial aspects (Great Circle sailing, composite sailing or plane sailing)
- Expected traffic situation during the voyage (COLREGS)
- Reporting Systems during the voyage
- Climatologically planning of the voyage (Passing of different climate zone which can affect the cargo, upwelling cold water)
- Meteorological voyage planning with regards to different sea-water temperatures, atmospheric temperatures and pressure, wind and current affecting the ship during the voyage and the expected sea state)
- Dynamic Stresses acting during the voyage and its influence on the cargo and ships integrity.
- Expected Vibration for the intended voyage (Stress acting due to vibration and possible shifting of cargo due to vibration)
- Stability and possible shifting of cargo
- Cargo Maintenance during the voyage _ with regards to weather condition and properties of cargoes.(Ventilation etc.)
- Ship resistance - mostly frictional resistance - expected during the voyage (SEEMP - Ship Energy Efficiency Management Plan)
- Fuel Oil consumption - SEEMP
- Minimum passing distance to shore or islands
- Minimum required water depth - important for the passage plan from berth to Pilot and wise versa - squat effect,

interaction with other ships and counter measurements for banking effect (especially the last one is mostly underestimated but due to the full hull form and therefore greater inertia of the Bulk Carrier compared to a Heavy Lift ship or Container Feeder service Ship, this is important not to run aground.)
- Ballast water management
- Ship maintenance during the voyage
- Limiting of voyage – load line zones

If these risks will be correctly determined and assessed the risk of loss or damage of cargo and or structural damages to the ship can be drastically reduced.

The master should be always reminded that a voyage plan is a prima facie evidence.

The principle should be explained on hand of an example. The example includes only the risk assessment, not the complete voyage plan. The first part is just an example of a voyage plan after the risk assessment was conducted. The second example deals with the risk assessment for an intended voyage from Gotenborg to Savannah with a consignment of steel coils, loaded on a five hatch bulk carrier.

Example for a final voyage plan

Master	Jens Nobody		General Information		Voyage Number		1208 - 15	
Ship Name	MV Amalia Jager	Date of Departure	08.08.2015	Time of Departure	08:00	UTC	15:00	
P.O.L	Seattle	Time Zone	UTC -7 hrs(DLST)	Displacement	19071.70 mt	Voyage Speed	13.5 knots	
P.O.D	Dutch Harbor	Time Zone	UTC -7 hrs.(DLST)	Total Distance	1600.00 nm	☑ Slow Steaming	Tides	
ETA P.O.D		Draught		ETCP[Days]	5.2	☐ Normal Steaming	Seattle	0.50 m
							Dt.Harbor	-0.05 m
Date	13.08.2015	Departure		Pre - Calculated EEOI	1.6	Cargo	Container/GC	
		Fwd	8.55 m				16230.00 mt	
		Aft	8.65 m					
Time[Local Time]	13:00	Arrival		EEOI which should not be exceeded		☑ Voyage Cold to Warm	☑ Voyage Warm to Cold	
		Fwd	8.45 m		3.5			
		Aft	8.55 m					

Paper Charts to be used	ECDIS Charts to be used	Admiralty List of Radio Sign.	Tide Tables and Tidal Atlas	Admiralty List of Lights /Fog	Admiralty Sailing Direction	Pilot Chart other Charts
Chart No according to Admiralty Chart Catalogue 47, 50, 46 4950, 4947, 4947 4945, 4944, 4920 4921.4978,4979 4969	According to Amiralty Chart Catalogue: Area 22	According to Admiralty Chart Catalogue: NP283(4) ADRS Vol6: Area2 ADRS Vol.1,3,4,5 Area 2 ADRS 1345	According to Admiralty Chart Catalogue: NP 206 - 16 Area 8	According to Admiralty Chart Catalogue NP 80 Digital Light List Area 8	According to Admiralty Chart Catalogue NP 25 NP 26 NP 4 NP 23	World Time Zones 5006 Pilot Chart for North Pacific, Month August, 5172(8) Mercator Sailing Chart 48°N to 60°N Ocean Passage of the World NP 136 Distance Table Pacific Ocean NP 305(3)

WP No or Leg	Latitude	Longitude	Course	Distance [nm]	Speed	Time to Go	Current	Wind	Fix Frequency	Remarks
1	Departure Seattle	N N	Various Courses	Tide during Departure: Low Tide : 0.5 m (see table in appendix)						2 hrs. Prior departure Prepare Bridge and engine. Engine on MDO, ECA Zone Re check entries in AIS and ECDIS system Print out of Voyage Plan
2	Pilot Stat Port Angeles 48° 15' 00" N	122° 22' 00" W		69.00 nm B.O.SP Special Note : UKC = 1.0 m all the time. Squat at Full Maneuver: 0.5 m After unberthing and tugs are off, clear lines fw and aft Full Bridge Team until end of Juan de Fuca Srait Predicted area of Risk Unberthing and Piloting Entering Seattle fairway - in and out bond traffic. Crossing of fairies - nearly the whole passage from berth to Pilot station Pilot Ground - In and out bound to Seattle and Vancover - dense Traffic From Pilot Ground to end of J.d.Fuca Strait - dense traffci due to sport fisher	12.50 knots	5.5 hrs.	Various	Various	3 min	Expected condition Dep. Windage are: 2100m² Underwater area: 1177m² Expected Wind: SE 10 knt Current: W 1.2 knt Force created by Wind and Current: 30 tons Required Bollard Pull: 30 tons Draught on Dept. 8.5 m One tug: Bollard Pull 38 tons all the time Master-Pilot exchange see attached list
3	48° 15' 00" N	123° 30' 00" W	274°	20.00 nm	13.50 knots	1.5 hrs.		W /BFT 2	3 min	Standing order of OOW Inform Master immideat if vessel is off of intended Track
4	48° 16' 00" N	123° 52' 00" W	290°	15.00 nm	13.50 knots	1.1 hrs.	Variuos	W /BFT 2	3 min	Bridge Team: Master Pilot, OOW Helmsman After unberthing Lookout on the bridge Prior departure. Take weather chart 500 mb and surface
5	48° 31' 00" N	124° 24' 00" W	272°	8.00 nm	13.50 knots	0.6 hrs.		W /BFT 2	3 min	
6	48° 28' 00" N	124° 39' 00" W		Info Boxes in sea chart: For transit message and communicaiton with VTS Juan de Fuca Strait use infor boxes in sea chart						

Example II - Voyage Plan Risk Assessment

Voyage: Gotenborg to Savannah
Cargo: Steel Coil – 22.000 tons, according to draught survey
Displacement: 34.340 mt
Draught: Fwd: 9.85 m   Aft. 9.88 m – Winter draft
L.O.A: 185,50 m
B.O.A : 32.00 m
Time of departure: 03.December
Average speed: 13,5 knots – Slow steaming
Voyage: Cold to Warm

Condition on Departure:
$GM_{corrected}$: **2.05 m**
Statical and Dynamic Stability: All IMO criteria fulfilled
Wind heeling criteria: Fulfilled
EEOI for the Voyage at 13,5 knots: 1,3. Company requirement: 3,5 - fulfilled
Expected Vibration: **43 vibr./min**
Expected Ship resistance: **385,3 kN at 9,85 m fwd draught and 9,86 m mean draught**
Pivot Point: **90,71 m from aft**
Propeller wash on departure and arrival at
- **Slow ahead / astern**   =   **0,50 m**
- **Half ahead / astern**   =   **0,70 m**

Required Bollard Pull: **65 mt** for an expected wind speed of 15 knots and current of 0,5 knots
Counter measurements for banking effect (P/Stb. Banking effect:
8° Port or Stb. Rudder

In General:

All additional information, like nautical publication to be used or Reporting Systems etc. are parts of the column "Remarks" or "Pilot charts and others" in the voyage plan and must be entered accordingly. Important information to be marked in the sea charts as well as No Go Areas, contingency anchorages etc. (See example given)

The risk assessment is the fundament of a successful voyage planning and must be completed prior finalizing the voyage plan.

Risk assessment for the above said voyage from Goteborg to Savannah – Five Hatch Bulk Carrier with Gears – loaded with steel coils

A. Risk Assessment Departure Goteborg – Pilot on board

Sheet of Risk Assessment		Voyage No:	
Passage Plan Sector: From Goteborg		To Savannah	

Perceived Hazard or Risk
(descirption of Risks which might occur)

Departure Gotenburg

1. Pilot unknown
2. Master - Pilot exchange
3. Tugs to be used
4. Required Bollard Pull
5. ECA Zones - Departure until end of English Channel
6. Propeller wash during departure maneuver
7. Wind and current situation on departure
8. Squat effect and Banking effect
9. Interaction between ships

8. Underkeel clearance and water depth
9. Communication with Pilot and Tugs

Risk Assessment(Tick Level)		Substantial			
Hazard Severity				**Likelihood of Occurence**	
	5 Very High	Mulitplied by		5 Very High	Risk Factor
4	4 High		4	4 High	Hazard Risk x Likelihood
	3 Moderate	16		3 Moderate	
	2 Slight			2 Slight	
	1 NIL			1 NIL	

Control Measuremnts:
1. Ship must be seaworthy. 2. Discuss the passage plan with the pilot in detail. Did not allow superficially explanation 3. Lateral surface area and lateral underwater area must be part of passage plan - influence on wind and current on the maneuverability of the ship. 3. Build Bridge team, Pilot nad Tugs are part of the team. 4. Check and control movements of tugs. Officer of maneuverstation should report the movements and if comply with the intention. 5. Toolbox meeting with officers, engineers and crews. 6. Ship remains on Diesel oil the whole transit from Goteborg to entrance in Northatlantic 7. Propeller wash max. = 0,70 m. Squat at 8 knots = 0,50 m astern. UKC required: 0,80 m. R.O.T max: 0,5

Results after Recommendations implemented

Efforts should be made to reduce the risk, but the costs of prevention may be taken into account

Hazard Severity		Likelihood of Occurence			
	5 Very High	Mulitplied by		5 Very High	Risk Factor
	4 High	**Moderate**	4	4 High	Hazard Risk x Likelihood
3	3 Moderate			3 Moderate	
	2 Slight	**12**		2 Slight	
	1 NIL			1 NIL	

Remarks

After discussing the passage plan and the departure manuever with the pilot, master must inform all officers / engineers and crew if there are any changes. If there are changes these changes must be implemented in the new passage plan
Frequently monitoring of the intended track and position of the ship by the OOW and Master. All orders given by the pilot to be repeated by the helmsman.
ECDIS should by only used aas an AID. OOW must make sure that position fix will be done by optical and radar bearings.
Parellel Indexing to be used. Communication in English only

B. Risk assessment Cargo - Cargo Maintenance during the Voyage

	Sheet of Risk Assessment	Voyage No: L 36a
Passage Plan Sector:	From Gotenburg	To___Savannah_____

Perceived Hazard or Risk - Risk assessment Cargo
(desciprtion of Risks which might occur)

1. Cargo Hold Temp. After loading about 3-4°C.
2. Max Humidity accepted: 52%.
3. Critical point: 09.december for intended route. Upwelling cold water from the Labrador current. Risk of ship sweat
4. Voyage os cold to warm
5. Temperature increase without ventilation: 13,3 days *0,3° = 4°. End temperature in Savannah = about 8°
6. Risk of sweat if opening the hatches in Savannah: Ait Temp: 18°C and cargo hold temp. 8°C
7. Ventilation is not essential, if ventilating risk of increase in humidity

Risk Assessment(Tick Level)   **MODERATE**

Hazard Severity				Likelihood of Occurence	
	5 Very High	Mulitplied by		5 Very High	Risk Factor
4	4 High			4 High	Hazard Risk x Likelihood
	3 Moderate		3	3 Moderate	
	2 Slight	**12**		2 Slight	
	1 NIL			1 NIL	

Control Measuremnts:
Efforts should be made to reduce the risk, but the costs of prevention may be taken into account. Risk reduction measures should be implemented within a defined time period. Where the moderate risk is associated with extremely harmful consequences,further assessment may be necessary to establish more precisely likelihood of harm as a basis for determining the need for improved control measures
If necessary alter course on the 4th day, 08.december to avoid the cold upwelloing water and therefore ships sweat
Daily preparation of cargo hold met.diagram. Monitoring of humiditiy. No ventilation - ventilation on masters order

Results after Recommendations implemented

Hazard Severity		Likelihood of Occurence			
	5 Very High	Mulitplied by		5 Very High	Risk Factor
	4 High			4 High	Hazard Risk x Likelihood
3	3 Moderate	9	3	3 Moderate	
	2 Slight	Tolerable		2 Slight	
	1 NIL			1 NIL	

Remarks

Monitoring is required to ensure that the controls are maintained
Daily checking of dew point inside the hatch and of atmospheric temperature
Each watch must take a weather chart - focusing on: Temperature charts
(sea and air temp. and wave height chart)
If necessary (weather wise) use alternative route to avoid damages
to cargo due to strong rolling, pitching and due to increase of humidity above 52%

If hatches will be opended in Savannah there is a temporary increase in humidity due to temperature differential -Hatch / outside air temperature (8°C/18°C)
Any claim is not justified, because the sweat will dry immidietly and accodring to charter party this is acceptable.

C. Detailed pre-analysis of ventilation requirement during the voyage

The steel coils should not exceed a maximum RH of 52% during the voyage
Condition on departure Gotenburg

RH- Air	Air Temp.	Dew Point Outside Air
56	3	1,9

RH Cargo	Temp.Cargo	Dew PointCargo
50	3	1,7

Assuming that the hatch is a semi hermetic closed body the temperature increase on this voyage/day is about 0,3°C/day without any ventilation. The voyage is from cold to warm – ventilation is not essential.

Expected condition, 05. December, 1800 lt, After passing English channel:

RH- Air	Air Temp.	Dew Point Outside Air
70	9	7,1

RH Cargo	Temp.Cargo	Dew PointCargo
50	3,8	2,2

No ventilation, the dew point of the cargo is less than the dew point of the atmospheric air temperature

Expected condition 0n the 9th December 1200 hrs. lt:

RH- Air	Air Temp.	Dew Point Outside Air
78	9,5	8,4

RH Cargo	Temp.Cargo	Dew PointCargo
51	4,8	2,8

No ventilation, the dew point of the cargo is less than the dew point of the atmospheric air temperature

Expected condition on the 11.December

At about 1200 hrs.lt, passing the Labrador current – cold water current
Air temperature forecasted (Pilot charts) with 2°C and sea surface temperature with 3,2°C

RH- Air	Air Temp.	Dew Point Outside Air
72	2	1,6

RH Cargo	Temp.Cargo	Dew PointCargo
51	5,4	3,1

Critical phase of the voyage because the cold upwelling water of the Labrador current might lead to ship sweat. Cargo Hold condition must be checked. If necessary course must be adjusted on the 8th / 9th December at Noon time, to avoid the ships sweat – see below

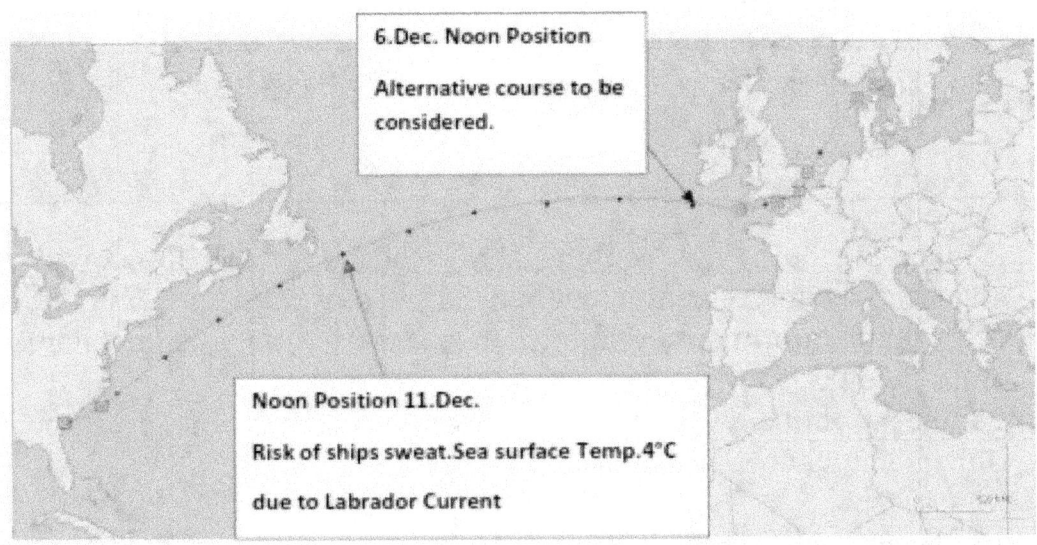

Alternative Route

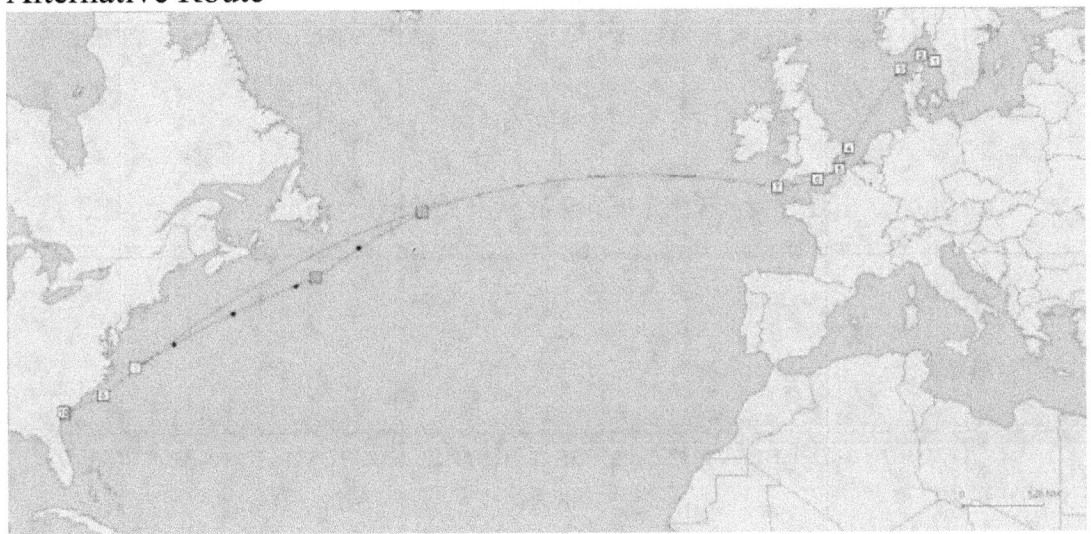

Condition on alternative route:

RH- Air	Air Temp.	Dew Point Outside Air
72	4	3,3

RH Cargo	Temp.Cargo	Dew PointCargo
51	5,4	3,1

If following the alternative route, ships seat and or cargo sweat will not occur. Course of action required. Check sea water temperature every 4 hrs. Inform master immediately if the sea water temperature drops below 3°C.
For the proof of any radiation – energy transferred the Stefan-Boltzmann Law was used.

16th December – arrival Savannah

RH- Air	Air Temp.	Dew Point Outside Air
82	17	15,8

RH Cargo	Temp.Cargo	Dew PointCargo
51	7	4,0

It might be that there will be sweat water formation as soon as the hatches will be open. This is not critical and is known by the receiver. No claim

Load line Zones:

Following load lines will be passed during the voyage – see chart below:
- North Atlantic Winter Seasonal Zone I
- North Atlantic Winter Seasonal Zone II
- Summer Zone – Approaches and Arrival Savannah

Ship can be loaded up to Winter North Atlantic

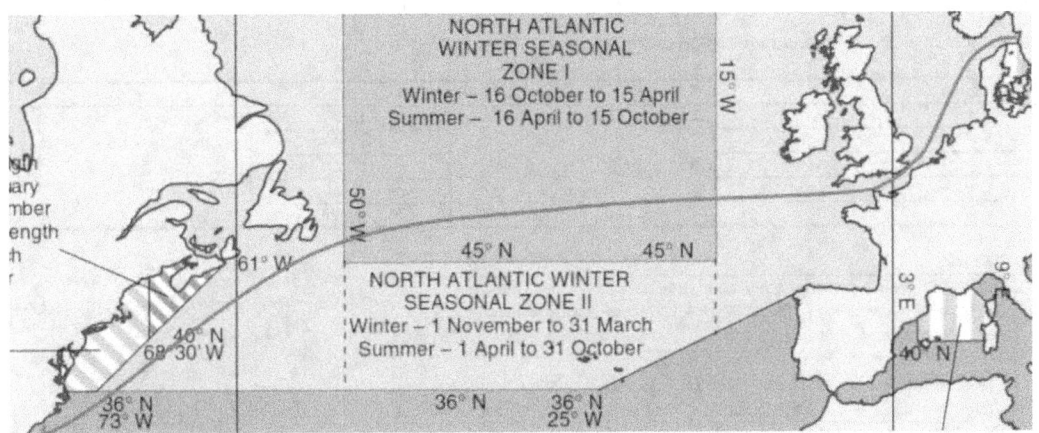

D. <u>Meteorological – Climatological Risk Assessment for the Intended Voyage</u>

	Sheet of Risk Assessment	Voyage No: L 36a
Passage Plan Sector:	From Gotenburg	To___Savannah_____

Perceived Hazard or Risk - Risk assessment Weather and Climate
1. Departure Gotenburg and passing the Baltic and North Sea up to German Bight: Humid Continental Climate and influences by the cold polar climate from the north. English Channel humid continental climate
2. Transatlantic passage - severe weather condition expected in the northern part of the North Atlantic
3. Climatological risk if passing from polar to humid continental cliamte - cargo in risk of increase in humidity
4. Cold upwelling water if passing the Labrador current
5. Warm humid climate if sailing along the east coast of US - influence of Gulf Current
6. Wave heights up to 4-5 meter
7. Strong Pitching and rolling can be expcted
8. sea sloshing - green water on deck -, pounding and panting

Risk Assessment(Tick Level) Substantial

Hazard Severity			Likelihood of Occurence	
	5 Very High	Mulitplied by	5 Very High	Risk Factor
4	4 High		5 4 High	Hazard Risk x Likelihood
	3 Moderate		3 Moderate	
	2 Slight	16	2 Slight	
	1 NIL		1 NIL	

Control Measuremnts:
Voyge should not be started until the risk has been reduced. Considerable resources may have to be allocated to reduce the risk.Where the risk involves work in progress,urgent actions should be taken
1. Weather reports every 6 hrs. after departure Gotenburg. 2. Check Air pressure frequently. Report any anomaly to master immediatley
3. Alternative course if necessary. Wind force will remain on alternative route, but wave height is reduced. If necessary speed to be adjusted accordingly. 4. Daily checking of hatch cover sealings. All doors,opening to be water tight closed

Results after Recommendations implemented

Hazard Severity		Likelihood of Occurence			
	5 Very High	Mulitplied by		5 Very High	Risk Factor
	4 High			4 High	Hazard Risk x Likelihood
3	3 Moderate	9	3	3 Moderate	
	2 Slight	Tolerable		2 Slight	
	1 NIL			1 NIL	

Remarks

If weather condition will be unchanged, use alternative route to avoid the more sever weather in the north. If low pressure moves more to the north, the intended route can be used. Final descion after passing Falmouth

Wave height should not exceed 3 meter. Disadvantage for alternative route: Wave direction against ships heading, therefrore speed to be adjusted accordingly. Increa:
Advantage: No strong rolling and if speed adjusted moderate pitching. All ventilation flaps to be closed.

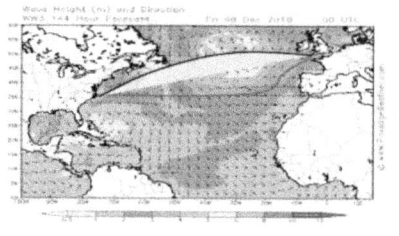

From Falmouth Transatlantic Passage to US East Coast:
Polar Climate and humid Continental climate.

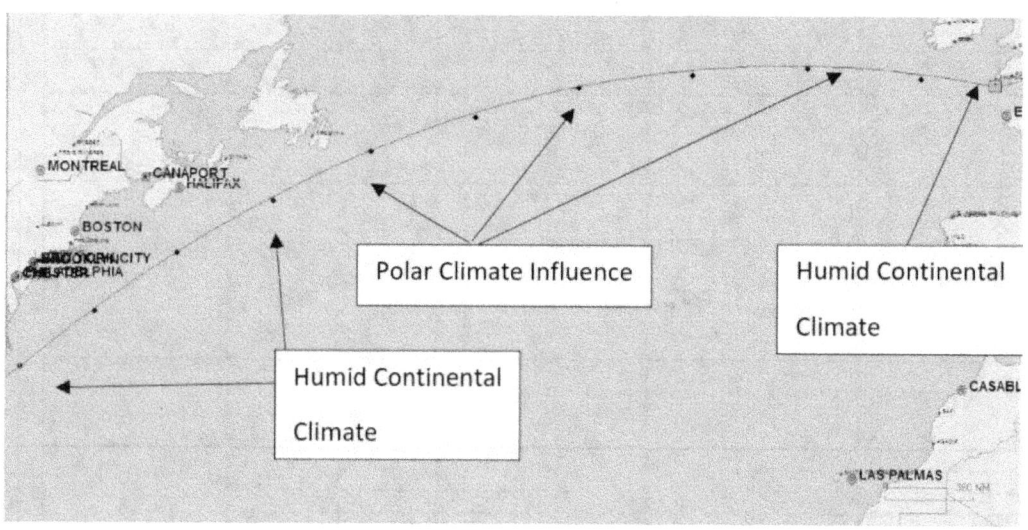

If passing Halifax until passing Norfolk the climate is a humid continental climate.

After passing Norfolk until arrival Savannah the climate is a humid subtropical climate. (Influenced by the Gulf Current), therefore also increase of sea water temperature during the last day passage from 15°C to maximum 23°C.

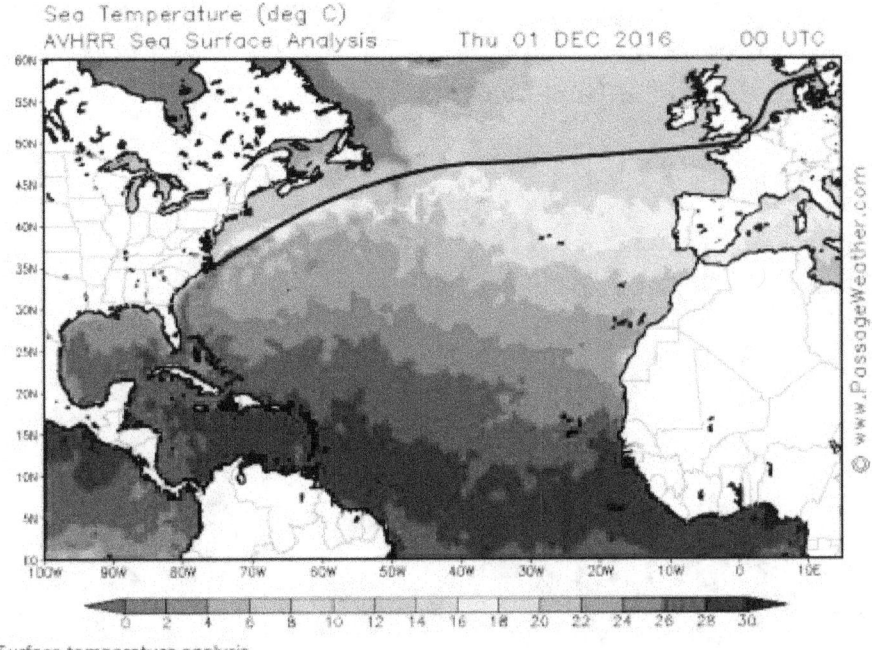

Surface temperature analysis

Due to the climate zones passing the ventilation of the cargo is restricted and should be only conducted if the cargo hold meteorological diagram presents the results for ventilating accordingly. If ventilating the cargo the humid air and later the humid subtropical air will enter the cargo hold and will increase the humidity of the hold drastically and above the maximum allowed 52%.

E. <u>**Meteorological Aspects to be considered during the voyage**</u>

The forecast analysis using the Pilot charts North Atlantic for the Month December presents Winds from West to NW'ly direction with maximum 30 knots (Bft 7-8) at about 50°N and 30°11'W with a maximum wave height of 4 m.

Current is setting in direction between 45° and 90° at the beginning until passing Argentia – US East Coast and abeam of St.John. Average: 0,6 knots

Afterwards current is setting in direction 180° and later in direction 225° abeam of Providence with an average of 0,5 knots

Than the current is setting in direction 45° (against the ship course) until arrival in Savannah with an average of 1,0 knots partly increasing to 1,7 knots (Influence of the Gulf Current)

Weather Prognosis using the forecast weather charts: Wind, Pressure and Wave height from 3rd December until 9th December (Forecast analysis done via internet resource -http://passageweather.com/ -

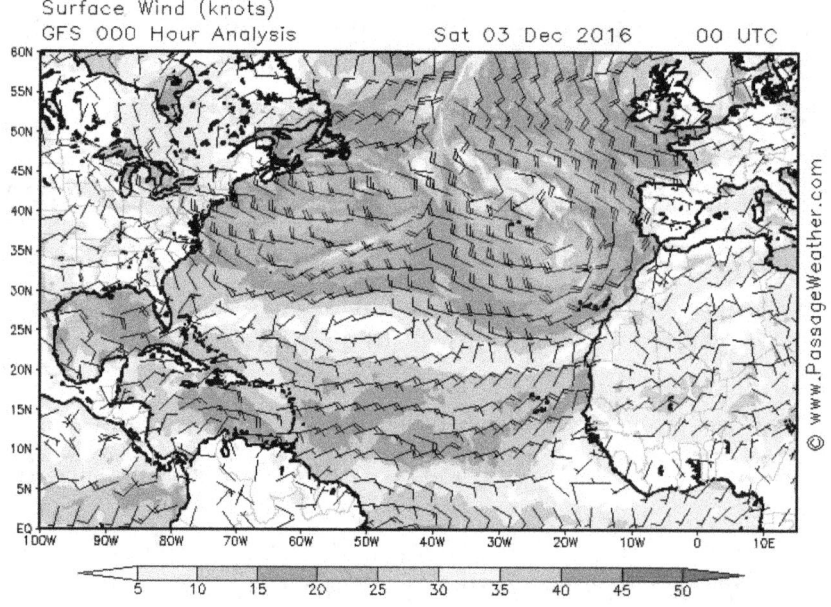

Surface Wind Analysis including intended track to Savannah

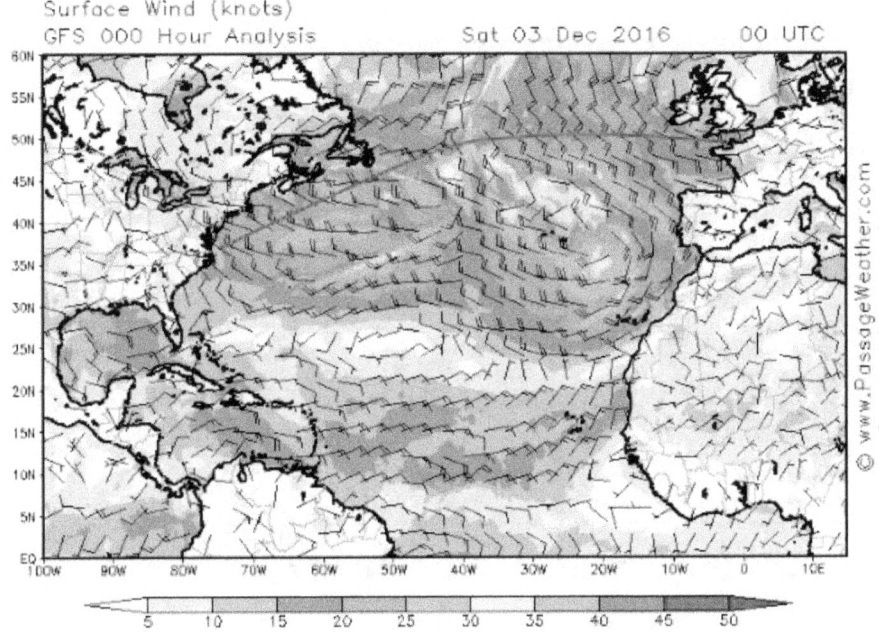

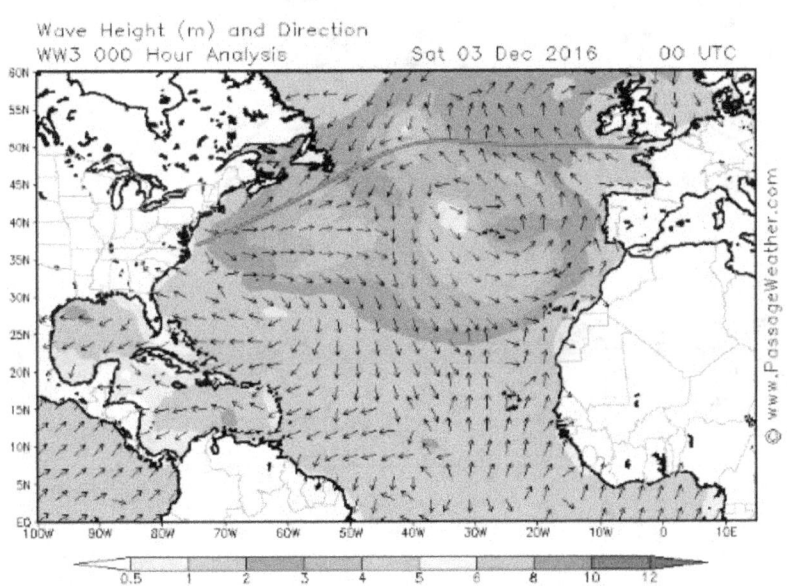

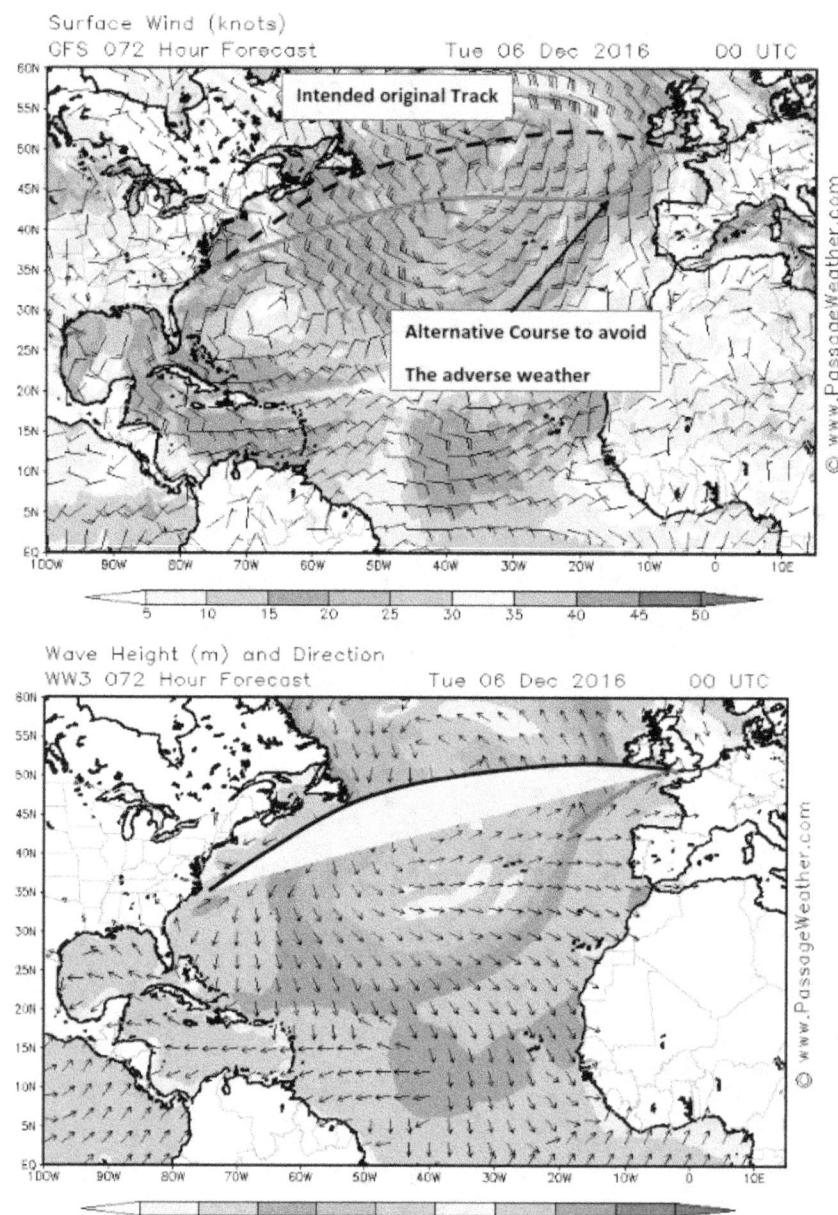

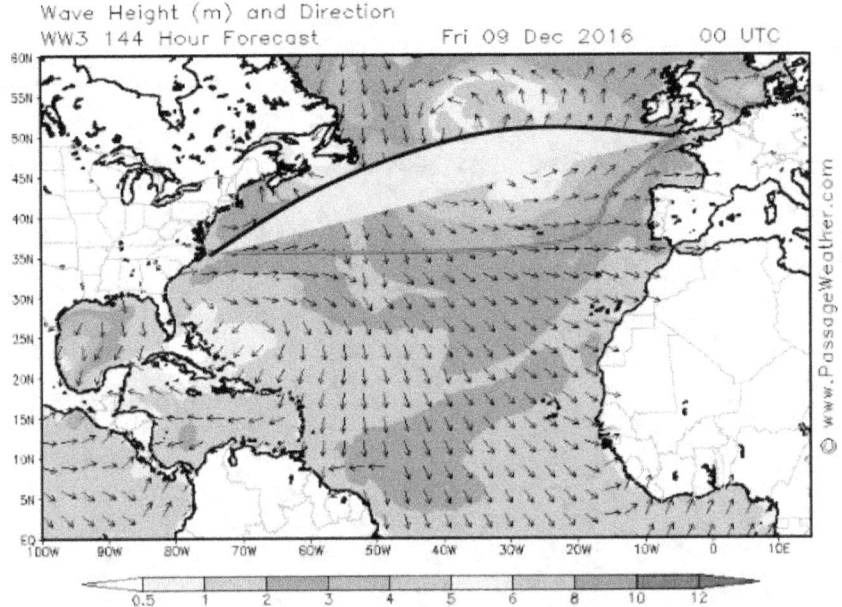

According to the analysis done, the intended course should be altered if the weather situation will be present as forecasted. The area to be avoided due to strong expected rolling and pitching is marked and the alternative route should be used (red marked). Vessel will mostly encounter head sea – 2-3 meter . The vessel speed must be adjusted accordingly but the risk of structural damages to ship and ort cargo is low compared if continuing the intended track. Therefore at the end of the English Channel a 48 hrs. weather prognosis to be taken for a final decision making.

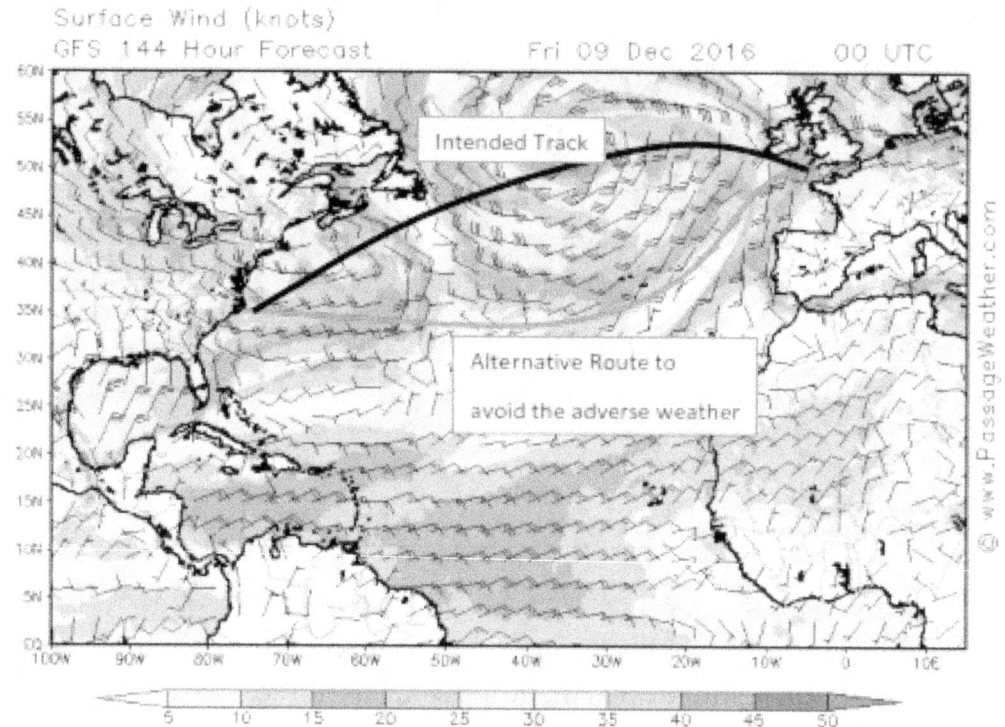

F. Risk Assessment Traffic Situation

Results after Recommendations implemented

Hazard Severity		Likelihood of Occurence			
	5 Very High	Mulitplied by		5 Very High	Risk Factor
	4 High			4 High	Hazard Risk x Likelihood
3	3 Moderate	9	3	3 Moderate	
	2 Slight	Tolerable		2 Slight	
	1 NIL			1 NIL	

Remarks
1. ECDIS is only a navigaitonal aid. Check position by using radar bearing and optical bearing
2. Be aware of the tidal current in the English Channel - check your average speed.
3. Keep a wide berth, if possible, if encountering fisherboats.
4. Brief lookout with regards to the high risk zones (Skagen, Dover Strait, English Channel, Approaches to Savannah.)
5. Do not be hazitated to change from autopilot to manual steering if the situation requires it.
6. Call master immediately if in doubt.

	Sheet of Risk Assessment	Voyage No: L 36a
Passage Plan Sector:	From Gotenburg	To___Savannah_____

Perceived Hazard or Risk - Risk assessment Ship's Traffic

Departure Gotenburg: In and outbound traffic and ferries. Fisherboats.
Passing Skagen: Fisherboats, small coasters , feeder service vessel. Inbound to Baltic Sea and out bound from Baltic Sea to North Sea
Passage English Channel: Dense traffic, corssing traffic passing Dover Strait. Fisherboats
Passing Casquetts: Crossing traffic from Cherbourg to Plymouth and Southampton. Fisherbaots, dense traffic and inbound vessel coming from Le Havre bound for the Gulf of Biscay and Northatlantic transit.
Passing Falmouth: Crossing traffic boun to Irish Sea. Be aware of small coasters. Inbound traffic from Northatlantic bound to North continent and port in the English Channel area.
North Atlantic Transit: Moderate ship traffic. Ship traffic gets more dense if apporaching the US East Coast - approaching Wilmington
Approches to Savannah: Traffic gets dense. In and outbound vessels bound for Savannah and Wilmington
Outbund traffic from Wilmington and Savannah to Jacksonville and Florida State Ports. Fisherboats.

Risk Assessment(Tick Level)		Substantial			
Hazard Severity				Likelihood of Occurence	
	5 Very High	Mulitplied by		5 Very High	Risk Factor
4	4 High		4	4 High	Hazard Risk x Likelihood
	3 Moderate	16		3 Moderate	
	2 Slight			2 Slight	
	1 NIL			1 NIL	

Control Measuremnts:
Voyge should not be started until the risk has been reduced. Considerable resources may have to be allocated to reduce the risk.Where the risk involves work in progress,urgent actions should be taken
1. Follow the COLREGS, especially Rule 10: Navigating in or near by Traffic Separation Zones. Always one lookout on the bridge.Master on the bridge if passing Skagen and Dover Strait
2. Master - Pilot Exchange regarding passage plan for Gotenburg and Savannah.
3. Adjust speed if necessary to avoid any accident - See COLREGS. Do not use ECDIS as a main navigational tool

6.0 Grain in Bulk

For grain there is special regulation. According to SOLAS Grain is: Maize, wheat grain, rice, wheat and some other grain products.

Grain has the ability to "float." 2% of the volume of grain will sack during the voyage. The average angle of repose is between 30° and 45°.

The ability to float and small angles of repose are the reason why the cargo shifts, if the vessel will face bad weather condition with long rolling periods.

There are strict regulations for loading grain. The humidity of the grain should not exceed 13% in winter time and 15% in summer time. So grain has to be always loaded dry. This has to be certified and the certificate has to be handed over to the ships command prior loading.

Also, the sacking of the grain during the voyage is negative for the vessel's stability. For this reason only vessels with an approved grain certificate can load grain.

6.1 Grain a sensitive cargo

Grain is a very sensitive cargo. Grain is sensitive against wetness and humidity[1]. Grain must be always loaded dry. If there will be rain during loading, the hatches have to be closed. Ventilation of the cargo is only allowed if the weather is good and it can be assured by the ship command that no spray-water will enter the hatches during ventilation. The whole hatch has to be free of any odour. This is also the main maintenance of the cargo during the voyage. It is from utmost importance that the ships command avoids all the above-mentioned criteria.

---

[1] The cargo can spoil

In most of the loading port, where grain products are shipped, the vessel will not commence loading until the surveyor has carried out his hatch survey and will certify that the vessel is clean and ready to load the grain product.

The table below presents some grain products and their S.F and density

Name	S.F	Density
Heavy Grain	48 cft/L^2 1.355 m³/mt	0,738 mt / m³
Milo	46,5 cft/lt 1,296 m³/ mt	0,773 mt / m³
Corn	47,5 cft/lt 1,324 m³/mt	0,755 mt/m³

Illustration 24 Grain Products and their Stowage factor (Source: P. Grunau)

6.2 Stability regulations and requirements if loading grain products

- It has to be assured that the cargo hold will be nearly filled
- If hatches are only partly filled, the cargo has to be trimmed even
- If necessary a centre bulkhead has to be build and set, that the cargo will not shift
- If necessary the unfilled space has to be filled up with bags of grain until the required stowage height is reached

Due to the fact that these calculation will exceed the normal procedures on board, classification societies require fixed Grain loading plans.

These loading plans must contain:

- Tables of volumetric heeling moments for the possibility that the grain product will shift in full holds[1], or a combination of common loading[2]
- Diagrams of the volumetric heeling moments, cargo volume, and centre of gravity in slack holds[3], accordance to the filling height.
- This will be assured in the plans and documents by calculating the angle of the grain surface with 25°heel[4]
- Calculated cargo loading examples with different S.F
- The improvement of these plans and documentation from the ship's command have to be presented to the Harbour master or authorities. The calculation must be done for the worst expected condition during the voyage. For small vessel[5] the requirement of calculation regarding SOLAS[6] is not valid

---

[1] This can happen if the grain product will sack during the voyage.

[2] Tweendeck and upper decks

[3] Partly filled holds

[4] Vol heeling moment / S.F = Mass heeling moment

[5] Under 500 BRT

[6] Safety of Life At Sea

7.0 Stability Related Problems on Bulk Carrier

Actually, most of the bulk carriers are facing longitudinal strength problems due to their construction. Stability problems will occur if cargo starts to shift, i.e. if the cargo has the tendency to float or to get liquefied (– Cargo which might tend to float is grain and cargo which might get liquefied is, for example, direct reduced iron ore). Here, the stability of the ship must be constantly observed during the voyage. If grain cargo is loaded, the cargo will settle and creates therefore an extra space between the top of the cargo and the top of the hold. The cargo can move which creates a list of the vessel resulting in a decrease of the stability of the vessel. If once the cargo starts to shift, other parts of the cargo will shift as well which can lead to capsize of the ship. The SOLAS Convention requires therefore that the upper ballast water tanks will be designed to prevent shifting. As well, the cargo should be trimmed and leveled. In the case of cargo which becomes liquefied, a shifting of the cargo will also take place. The reason is that fine concretes, if mixed with water, create mud at the tank top of the hatch. This mud can now easily shift and produce a free surface effect in the hatch. Shifting of cargo means always a list of the vessel with reduced stability. The shifted cargo will increase the forces acting on the structure inside the hatch. If the cargo will shift in longitudinal direction, the transverse bulk heads will be affected and the forces acting on the transverse bulk heads are also drastically increasing. Shifting of the cargo is the shift of the center of gravity mostly in the transverse direction, sometimes also in the vertical direction. A shifting of cargo goes alongside with a free surface which reduces the stability. The equation below represents the changing of the center of buoyancy for the vertical shift zB and for the transverse shift yB in regards to the momentum of inertia of the cargo hold I_{TO}

$$y_B = \frac{\tan\varphi \, x I_{TO}}{V_o} = \tan\varphi \, x\overline{BM}_o$$

$$z_B = \frac{\tan\varphi \, x I_{TO}}{2xV_o} = \frac{\tan^2\varphi}{2} xBM_o$$

Formula for the shifting of the center of buoyancy

Using this formula and the formula for getting the upright arm, the up righting arms can be calculated for each frame of each vessel.

$$for \, \varphi \leq 5°, \, h_{upright} = GM_o x \sin\varphi$$

$$for \, 5° < \varphi > 10°, \, h_{upright} = GM_o x \sin\varphi + z_B x \sin\varphi = (GM_o + \frac{\tan^2\varphi}{2} xBM_o) x \sin\varphi$$

$$for \, \varphi > 10°, \, h_{upright} = GM_o x \sin\varphi + LK_L - KM_o x \sin\varphi$$

$$where: LK_L = Kb_o x \sin\varphi$$

We can now conclude that the floating ability of bulk cargoes is only in relation to its angle of repose (see also chapter: Angle of Repose), and if the angle of heel φ is greater than the angle of repose α plus 7°, the cargo will shift. The maximum angle of heel at grain cargo (International Grain Code) is 12°.

Now it can be concluded :

1) $\varphi > \alpha + 7° = shifting \, of \, cargo$

2) $h_{upright} - h_{heel} > 0 = stable$

3) $h_{upright} - h_{heel} = 0 = indifferent$

4) $h_{upright} - h_{heel} < 0 = instable, capzi\sin g$

The stability can now be calculated by calculating the shifted center of gravity to get the transverse shift of the center of gravity and by calculating the heeling arm h_{heel}. If only the angle of heel is known, the remaining stability can be solved graphically. A perpendicular line will be drawn in the stability curve which will intersect the static curve at the actual angle of heel.

During loading, discharging and the voyage, the structural integrity of the ship must always be in its limitations. If these limitations will be exceeded, the result is over-stressing of the structure, which means a drastic increase on the shear and bending moments. An overstress of the ship structure will cause additional dynamic loads acting on the ship which will have an influence on the longitudinal strength and also on the stability of the vessel.

The loads which are acting on the ship if floating in still water are static loads. These loads are:

- Actual load of the ship's structure including equipment and machinery

- Cargo loaded (weight)

- Bunkers and consumables

- Ballast water

- Hydrostatic pressure, sea water pressure acting on the hull

Additional to the static loads there are dynamic loads exerted on the ship's hull by sea and wave and the ships motion, like acceleration, slamming, and sloshing shocks. If now the ship will be overstressed, the static loads - static stress – will increase and will reduce the capability of the structure to sustain dynamic loads. Therefore, a bulk carrier is in the first case prone to longitudinal strength problems than to stability problems. But if the longitudinal strength is above the given limitations, the stability of the ship will be as well reduced. If a bulk carrier – and not only a bulk carrier, also all other ship types – will be loaded, the ship's lightweight and deadweight will be supported by global buoyancy up thrust acting on the exterior of the hull. This causes differences in the vertical up thrust along the ship's hull and the ship's weight, the downwards acting forces.

Therefore, these net vertical forces which are unbalanced cause the ship girder to shear and bend. A still water shear-force and bending moment will be created at each section of the hull. At sea, these cyclical shear and bending stresses will be additionally increased by dynamic stresses acting on the hull due to changing of wave pressure which is acting against the hull. To carry these loads, more longitudinal structural members are built in. During loading, there is additional local strength acting on the transverse bulk head, the double bottom, and the cross deck structure. Overloading will induce greater stress in the double bottom tanks, the hatch coamings and hatch corners, the transverse bulkheads. This can lead to an excessive structural reduction, causing also a reduction of the stability.

8.0 The Code of Safe Practice for Solid Bulk Cargoes

The dangers associated with the carriage of cargo in bulk have been known for a long time and the 1960 International Conference on the Safety of Life at Sea recommended that IMO draw up an international code of safe practice dealing with this subject. Work began immediately and in 1965 the first Code of Safe Practice for Solid Bulk Cargoes was adopted.
The Code has been updated at regular intervals since then and is kept under continuous review by the Sub-Committee on Containers and Cargoes. The practices contained in the Code are intended as

recommendations to Governments, ship operators and shipmasters. Its aim is to bring to the attention of those concerned an internationally-accepted method of dealing with the hazards to safety which may be encountered when carrying cargo in bulk. The Code does not deal with the transport of grain, which is covered by the International Code for the Safe Carriage of Grain (International Grain Code).

The Code of Safe Practice for Solid Bulk Cargoes deals with three basic types of cargo: those which may liquefy; materials which possess chemical hazards; and materials which fall into neither of these categories but many nevertheless pose some dangers, as stated above.

The Code highlights the dangers associated with the shipment of certain types of bulk cargoes; gives guidance on various procedures which should be adopted; lists typical products which are shipped in bulk; gives advice on their properties and how they should be handled; and describes various test procedures which should be employed to determine the characteristic cargo properties.

8.1 General precautions

It is of fundamental importance that bulk cargoes be properly distributed throughout the ship so that the structure is not overstressed and the ship has an adequate standard of stability.

Loaded conditions vary according to the density of the cargo carried. General cargo ships are normally constructed so that one ton of cargo occupies about 1.39-1.67 cubic metres of space when loaded to full bale cubic and deadweight capacity. The ratio of volume of cargo to its mass is known as the stowage factor. When the high density bulk cargoes with a stowage factor of about 0.56 cubic metres per ton or lower are carried, it is particularly important to pay attention to the distribution of weight in order to avoid excessive stresses on the structure of the ship. Since hull arrangements vary, it is not possible to establish overall rules applicable to all types of ships.

It is essential that the master be provided with loading information sufficiently comprehensive to enable him to load the ship without overstressing the structure. This applies to localized stresses on the structure as well as on the bending stresses. The master must also be able to calculate the stability of his ship for the anticipated worst conditions during the voyage. The initial transverse stability of a ship is usually expressed as the meta-centric height or GM. A large distance between the centre of gravity of ship and cargo (G) and the metacentre (M) means that the ship has adequate stability. As G approaches M, i.e. when the centre of gravity of ship and cargo rises, ships, when forced from a position of equilibrium, recover this position sluggishly.

Generally speaking, high density cargoes should be loaded in the lower hold spaces rather than the tween decks. Particular care should be taken when a ship has a high GM. In order to prevent cargoes from shifting the considerations dealt with below under the heading Bulk cargoes having an angle of repose greater than 35 degrees should also be taken into account.

The Code gives various precautions to be followed when information on the physical properties of the cargo is not available.

The Code lists other general precautions such as the need to protect machinery and the interior of the ship from dust and to ensure that bilges and service lines are in good order and not damaged during loading.

8.2 Bulk cargoes having an angle of repose less than or equal to 35 degrees

When a bulk cargo is emptied on to a flat surface, such as the hold of a ship, it forms a cone whose angle of repose varies according to the type of cargo. This angle is the one formed between the horizontal plane and the cone slope.

Cargoes with a low angle of repose are particularly liable to dry-surface movement aboard ship. To overcome this problem, the Code states that such cargoes should be trimmed reasonably level and spaces in which they are loaded should be filled as fully as is practicable, without resulting in excessive weight on the supporting structure

Special provisions should be made for stowing dry cargoes which flow very freely, in a similar manner to grain.

Securing arrangements, such as shifting boards or bins, should be used whenever the amount, location or properties of the cargo could cause excessive heeling through cargo shift, taking into account the density of the cargo.

8.3 Bulk cargoes having an angle of repose greater than 35 degrees

Generally speaking, high-density cargoes, such as most iron ores, have a high angle of repose, i.e. above 35 degrees.

The Code states that high density cargoes should be loaded entirely in the lower holds of the ship unless this results in the ship being too "stiff" or in the cargo weight on the bottom structure being excessive. It should be trimmed sufficiently level to cover evenly all of the tank top, to reduce the pile peak height and equalize weight distribution. In some circumstances the pile peak may be allowed to extend through the 'tween-deck hatchway but the Code says that the importance of trimming as a means of reducing the possibility of a shift of cargo can never be over-stressed. This is particularly true for smaller ships of less than 100 metres in length.

Trimming also helps to cut oxidation by reducing the surface area exposed to the atmosphere. It also helps to eliminate the "funnel" effect which in certain cargoes, such as direct reduced iron (DRI) and concentrates, can cause spontaneous combustion. This occurs when voids in the cargo enable hot gases to move upwards, at the same time sucking in fresh air. This effect is obviously not desirable, since it escalates the process of spontaneous combustion.

The Code goes on to list various considerations which should be taken into account when cargo is loaded in the 'tween-decks to reduce "stiffness".

8.4 Safety of personnel

After listing various regulations adopted by the International Labour Organisation, which should be taken into account during cargo handling operations, the Code gives details of other dangers which may exist. Some cargoes, for example, are liable to oxidation which may result in the reduction of the oxygen supply, the emission of toxic fumes and self-heating. Others may emit toxic fumes without oxidation or when wet. The shipper should inform the master of chemical hazards which may exist and the Code gives details of precautions which should be taken.

Health hazards can arise because of dust, and some cargoes can create dust or emit flammable gases which create a danger of explosion.

8.5 Cargoes which may liquefy

These include concentrates (materials obtained from a natural ore by a process of purification, by physical or chemical separation and removal or unwanted constituents), some coals and other materials with similar properties.

One purpose of this section of the Code is to draw attention to the latent risk of cargo shift and describe precautions which should be taken. Concentrates and similar finely-particulate materials may appear to be in a relatively dry granular state when loaded and yet may contain sufficient moisture to become fluid under the stimulus of compaction and vibration. In the resulting semi-fluid state, the cargo may flow to one side when the ship rolls but not completely return when the ship rolls the other way. As in the case of cargoes liable to shift, this can result in the ship reaching a dangerous heel or eventually capsizing.

The stability of the ship is also likely to be affected by "free surfaces"[5] of liquids in the cargo spaces. General cargo ships should only carry bulk cargoes which have a moisture content below the transportable moisture limit (TML), which is 90% of the flow moisture point (FMP) unless they are fitted with special arrangements to restrain the cargo.

Cargo ships in which internal structural boundaries are sufficient to limit cargo shift may also carry cargoes whose moisture content exceeds the transportable moisture limit. All ships which carry cargoes of this type should carry evidence of approval of the flag State. The Code stipulates the data which should be included in submission for approval.

To prevent possible increases in the liquid content of concentrates, cargoes containing liquids (other than canned goods or the like) should not be stowed in the same compartment as cargoes which may liquefy. Precautions should be taken to prevent water entering holds; this is even more important where contact with seawater could lead to serious corrosion problems for hull or machinery. In this connection masters should be aware of the possible danger of using water to cool combustible materials such as coal at sea, as this may well bring the moisture content to a flow state or create other hazards. Water, if used, is most effectively applied in the form of spray or mist.

8.5.1 Correct course of action if loading Cargoes which may liquefy

When the cargo is subject to recurring cycles or cyclic forces, such as the movement of the ship (rolling/pitching/slamming), the volume of spaces between the particles reduces, which causes the pore water pressure to rise, reducing the shear strength of the particles.
Pore water pressure refers to the pressure of water held within a soil or rock, in gaps between particles (pores). If the pore water pressure increases enough, the cargo can reach its flow moisture point.

The cargo enters a stage of transition whereby it begins to react like a fluid because of the loss of friction between the particles. This is called liquefaction.

What must be observed by the ship command prior loading these cargoes?

Sampling Procedure
Section 4.6 of the IMSBC Code outlines the Sampling Procedures for Concentrate Stockpiles:

• Consignments < 15,000 tones = 1 x 200g subsample each 125 tones
• Consignments > 15,000 tones < 60,000 tones = 1 x 200g sub-sample each 250 tones

- Consignments > 60,000 tones= 1 x 200 g subsample each 500 tones

Example:
A vessel is consigned to load 50,000mt of nickel ore
The IMSBC Code states
Consignments of more than 15,000t but not more than 60,000t; one 200g sub-sample is to be taken for each 250t to be shipped
In total 200 sub-samples should be taken
If the cargo certificate reads: **In total 180 sub-samples are taken, or in total 200 sub- sample - each 180 g,** than the cargo should not be accepted, because the shipper or charterer are not fulfilling the requirements of section 4.6 of the IMSBC.

Masters and officers should if possible undertake a visual inspection of the cargo before loading, to establish any parts of the consignment which may be appreciably different in moisture content. If this is the case, additional testing should be conducted to determine moisture content. Any parts of cargo found to be in excess of its transportable moisture limit should be rejected as being unfit for shipment.

A flow state is considered to have been reached when the moisture content and compaction of the sample produce a level of saturation such that plastic deformation occurs. At this stage, the moulded sides of the sample may deform, giving a convex or concave profile.

Signs of plastic deformation:
- *Moulded sides of sample may deform*
- *Cracks may develop on the surface*
- *Sample begins to show tendency to stick to bottom of mould*
- *Tracks of moisture on the table after testing*

The Master must ensure:

- *Must ensure the ship has received all the documentation necessary for the safe loading of the ship. The master should not begin loading until valid certification has been provided*
- *should not load any parcel of cargo which is in excess of its transportable moisture limit*
- *if there is any doubt as to the validity of the certificates for moisture content and transportable moisture limit, the master or his appointed representative should carry out tests before loading*
- *an independent surveyor can assist the master in sampling for moisture content analysis (which may be a local requirement)*
- *in tropical countries additional moisture content analysis may be necessary*
- *the 'can test' method alone should never be solely relied upon*
- *a visual inspection of the cargo will identify the stockpiles before loading to establish parts of the consignment which may have appreciable differences in moisture content*
- *ensure the cargo is retested for moisture content and*
- *transportable moisture limit if it is ' wetter' after rain exposure, or if the certificate is not correct*
- *ensure the cargo is loaded and trimmed as evenly as possible*
- *IMSBC Code must always be followed*
- *Resist any commercial pressure*
- *contact local P & I correspondent if master is in doubt of the suitability and safety of the cargo*

Example of calculation

The Moisture content of the cargo is 28%. The flow moisture point was measured with 21%.
Can you accept the cargo to be transported?

1. Calculate the TML = 90% of the FMP

$$TML = \frac{21\% * 90\%}{100\%} = 18,90\%$$

2. Compare the Moisture content with the TML

The moisture content is higher than the TML. The TML represents the maximum moisture in percentage where the cargo can be transported. If the MC > TML the cargo cannot be loaded

Example 2:

Your ship arrives in an Indian Port to load Iron Ore fine. Due to unavailability of the berth the ship have to stay at anchor for 8 days. Time of arrival at the anchorage: 21.04.2016.
On the 29th April in the morning the ship will be berthed. Upon arrival the shipper presents the cargo certificate which states that the laboratory test were carried out on the 20th of April 2016.

a. Can you accept the cargo for loading and what is your correct course of action?

The cargo cannot be accepted for loading, because the validity of the certificate is only seven (7) days, means the certificate is expired on the 27th April 2016. Your course of action is to require a new laboratory test which must be certified and must have the correct validation.

8.6 Mineral Concentrates and Ore in Bulk – Cargoes which might liquefy

8.6.1 Shifting of Mineral Concentrates and Ore in Bulk

Refer to the Merchant Shipping (Carriage of Cargoes) Regulations 1999, as amended. When carrying these types of cargo, in addition to complying with such regulations, the MCA for example has recommended that the Shipping Industry should comply with the IMO Code of Safe Practice for Solid Bulk Cargoes.
Cargoes like Iron Ore fines – concentrated, means cargoes which are chemical wise reduced, tend to be liquefied. These cargoes are listed in the IMSBC Code in group A.

8.7 Cargoes of Cargo Group A – Cargoes which may liquefy

8.7.1 Cargoes which may shift related to their internal friction characteristics

We can consider following subdivision:
The first group consists of those materials which are relatively dry and whose tendency to shift is related to their internal frictional characteristics as represented by the natural angle of repose. When the natural angle of repose of the material exceeds 35 degrees no specific recommendations are considered necessary apart from adequate trimming athwart ships

For cement special recommendations are given. If the loading of cement cargo is completed, the master should not immediately depart. Cement should first settle prior departure. Therefore it is recommended that the master will leave one to two hours after competition of the cargo, because the cement is settle and a shifting will most likely not happen.

When relatively dry cargoes are loaded in the lower holds only it is generally sufficient to trim out so as to fully cover the tank top and reduce the height of the peak. If cargoes are loaded in the tween decks it becomes necessary to trim the cargo in the lower holds to a much greater extent. The cargo in the tween decks should be trimmed to a reasonably level and should extend over the whole compartment. If the natural angle of repose is 35degrees or less particular attention should be paid to trimming and as many compartments as possible should be completely filled with cargo.
When it is known to the master and it is indicated in the Code and or description of the cargo received prior loading that shifting is probable, the need for shifting boards etc., should be judged against the ship's relevant intact stability characteristics.

For Example grain:
Grain has the ability to "float „and, therefore, also the ability to shift.
2% of the loaded volume of the grain will settle during the voyage.
The setting of shifting boards are very common.

8.7.2 Cargoes which might shift due to their moisture content

The second group consists of those materials in which moisture is held in free suspension between the very small individual particles.

If water is added to particles, water coating grains tend to bend them together by its surface tension. This gives a greater internal cohesion and, therefore, more shear strength

But if water is added to completely saturate the pore space, the pore water will act as a lubricant between the grains, and the pore pressure will force the grains apart from each other, resulting in less shear strength and angle of repose.

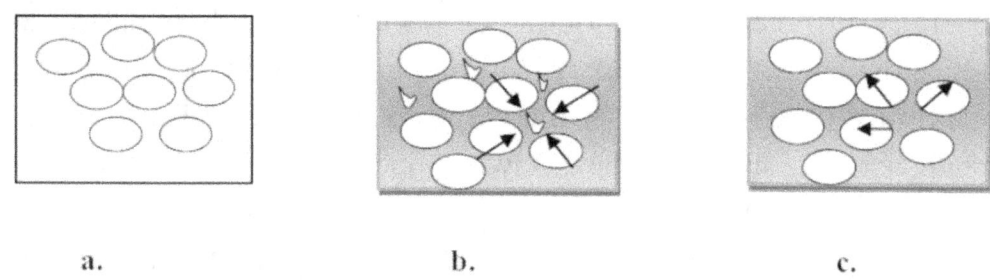

a.   b.   c.

Illustration P.Grunau – Cargo Handling and Stowage – BOD Verlag Norderstedt

Illustration (a) presents the situation when no water is added. The pores are filled with air. The particles will interlock with each other and the angle of repose is smaller than when the cargo is wet.

During the voyage the ship is facing several different movements and certain stimulus, like rolling, pitching and vibration. If there is too much water content and or moisture – free moisture content - inside the cargo, these moisture will be presses out and causes that the cargo will shift.

Cargoes of concentrate or similar materials of the group may be loaded in any type of ship without additional consideration of the adverse effects of the presence of free moisture provided the actual 'moisture content' of the cargo does not exceed the 'safe transportable moisture limit.' For such cargoes the recommendations made under the first group for dry cargoes will apply to their natural angles of repose.

8.7.3 Moisture content exceeding the safe transportable moisture limit

Where it is proposed to carry in specially suitable ships bulk cargoes having a 'moisture content' in excess of the 'safe transportable moisture limit' the decision if the cargo can be transported should be based on a survey carried out by a certified company or classification society and must be also agreed by the Headquarter of the shipping company.

The strength of the hull structure forming the boundaries of the cargo compartment, details of the intended cargo and the ship's intact stability characteristics in the anticipated loaded conditions should be examined. At the time of loading there should be available for the inspection of the Master and/or the Surveyor approving the loading, two certificates of 'moisture content'; one showing the 'critical moisture content' and the other showing the 'actual moisture content' of the proposed cargo. Normally the required information are one certificate which indicates the flow moisture point – critical moisture content, the transportable moisture point (TMP) of the critical moisture content, which is equal to 90% of the Flow moisture point (FMP).

Further the certificate indicates the moisture content of the cargo. If these information, besides a lot of others, are not on hand and the validity of the certificate is older than seven days, the cargo cannot be accepted for transportation.

8.7.4 Where shifting of cargo is probable

Shifting of cargo is problem if the angle of repose of the cargo is 35° or less or where the moisture content exceeds the 'safe transportable limit'.
If this is the case the intact stability of the ship may be in risk. Further the structural integrity of the ship will be reduced – see Iron Ore fines or reduced Iron Ore.

Consideration for the intact stability of the ship

The Intact stability of the ship may be considered adequate if, after taking account of any cargo shift, the following obtains:
The angle of heel does not exceed 65

$$\text{Rise of } KG = \frac{Total\ vertical\ shifting\ moments}{Displacement}$$

> Example: The deck edge immersion or also called inflexion– see illustration - will take place at 15°

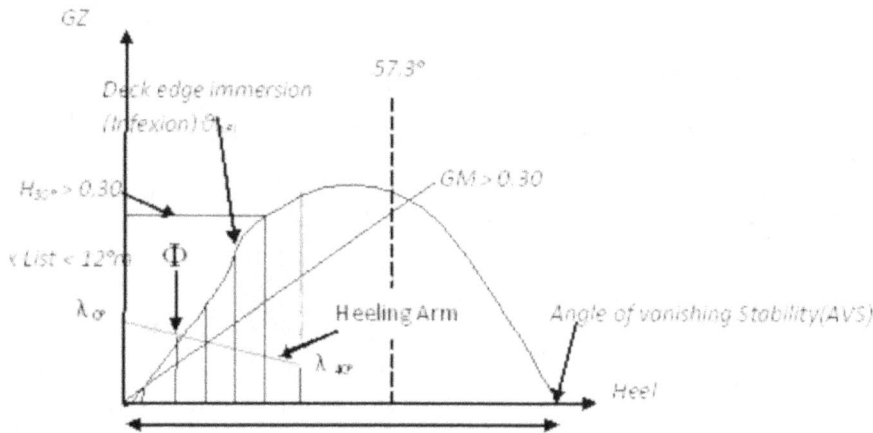

Illustration 25 Deck edge Immersion – inflexion– Source: P.Grunau

Therefore: 65% of 15° = 9.75°
   15°+9,75° = 24,75°.
   The angle of 24.75° should not be exceeded if cargo shifts
   ➤ The residual dynamic stability, measured up to 30 degrees beyond the angle of heel, see illustration No 3 – yellow shaded area - is not less than 0.10 meter-radians.

8.7.5 Calculation of the cargo shift moments and the reduction in KG

The cargo shift moments for anyone continuous section of the hold should be calculated: (MSIS003/Part 8/Rev.1.02 – Load Line Instruction):

$$\text{horizontal heeling moment} = \frac{1}{12} p \tan^2 \alpha \int_0^l b^3$$

$$\text{*Vertical moment} = \frac{1}{12} p \tan^2 \alpha \int_0^l b^3$$

Where l = length of section of hold

b = breadth of section of hold

p = density of cargo

α = surface angle shift (this to be taken as 35 degrees minus the natural angle of repose or 20 degrees in the case of cargoes which exceed the safe transportable moisture limit).

* This value divided by the ship's displacement will give the resultant rise in the ship's KG

The calculation will give us the horizontal shifting moments and the vertical shifting moment, means the horizontal shifting moments will

90

be used to get the later list of the ship and the vertical shifting moments will be used to calculate the rise in the ship's KG and therefore represents the reduction in GM (KM-KG = GM_{solid}) and $GM_{(solid)}$ – free surfaces = GM_{fluid}.

An example will proof the results and the reduction in GM

MV" M. Chripet," a seven hatch Bulk carrier, is loaded with Iron concentrates. Total displacement = 56.000 mt

The cargo shifts in all hatches, assuming that the breadth and length of the section are identically.

The length of the section of the hatch is assumed with 16 m and the breadth 25 m.

The natural angle of repose for this cargo was given with 33°. Density of the cargo is 2.3 t/m³

What will be my new KG if the KG prior shifting of the cargo was 13.40 m and the KM was 14.30 m?

What are my shifting moments /hatch and in total

Step I:
Given: L = 16 m; b = 25 m
         p = 2.3 t/m³
         α = 35°-33° = 2°
         Δ = 56.000 mt

Step II:
Formula for Horizontal shift

$$Horizontal\ shift = \frac{1}{12} * p * tan^2\alpha \int_0^l b^3$$

Step III:
Dissolve the integral of the given formula for calculation

$$Horizontal\ shift = \frac{1}{12} * p * tan^2\alpha \int_0^l b^3$$

$$Horitontal\ shift = \frac{p * tan^2\alpha}{12} * \frac{L * (b^4)}{4}$$

Step IV:
Substitute the given values into the formula and solve for the horizontal shifting moments

$$Horitontal\ shift = \frac{\frac{2.3t}{m^3} * tan^2 2°}{12} * \frac{16m * (25^4 m)}{4}$$

$$Horizontal\ shift = 2.34 * 10^{-4} * 1562500 m^4$$

$$Horizontal\ shift = 365{,}625 m^4\ per\ hatch$$

Step V:
We assumed that the section of each hatch is identically, therefore the total shifting moments for all hatch are: $365{,}625 m^4 * 5$ hatches = $1828{,}125 m^4$

Step VI:
Calculation of the vertical shifting moments and the rise KG
The vertical shifting moment calculation is identically with the horizontal shifting moment calculation, therefore the result is the same.
To get the new KG, the total amount of all shifting moments have to be divided by the displacement.

$$Rise\ of\ KG = \frac{Total\ vertical\ shifting\ moments}{Displacement}$$

$$\text{Rise of } KG = \frac{1828{,}125 m^4}{{'}56.000\ mt} = 0.03\ m$$

New KG = Old KG + change = 13.40 m + 0.03m = 13.43 m
New GM = 14.30 m – 13.43 m = 0.87 m

8.7.5.1 Example Calculation to justify the necessity for this calculation

Another example will emphasize the drastically changes. For this example the displacement remains constant, length and breadth of section as well. The TML is above the MC of the cargo - means acceptable. The density of the cargo remains constant, too

Changes will be done for the natural angle of repose only. The consequence is an increase of the angle α, therefore an increase in the nominator with a constant denominator which will automatically lead to an increase of the horizontal and vertical moments and therefore also of a more rise of the KG - assuming that all other parameters remain constant for all the examples.

Example I - Angle of natural repose = 33°

Calculation of Shifting Moments and Rise of KG for shifted Mineral Concentrates

Flow Moisture Point	28%	Transportable Moisture Limit		25%	Moisture Content Cargo	23%
Density of Cargo [t/m³]	Angle of Repose [°]	Section Length [m]	Section breadth [m]	Angle α	Displacement [mt]	No. Of Hatches
2,30	33,0	32,00	30,00	2,0	56000,000	5

Horizontal shifting moment	7572,85		Final List after cargo shifted[°]	
Vertical Shifting Moments	7572,85		4,5	
Rise of KG	0,135		Deck edge immersion in still water	16
			Allowed increase: 65%	10,4
Old KG	9,5		New deck edge Immersion	26,4
New KG	9,635		Consequence if Cargo shifted	
KM prior shifting	10,73		No risk but correction required	
Final GM	1,09		Pre-requisite Ratio TML/MC	
			Cargo can be transported	
			Accept Cargo	

The Final list if the cargo shifted in all hatches is 4.5°, which is below the deck edge immersion and the new allowed deck edge immersion (should not exceed 65% of the deck edge immersion in still water). Also the GM will be reduced but is still acceptable.

Example II - Angle of repose = 31°

Calculation of Shifting Moments and Rise of KG for shifted Mineral Concentrates

Flow Moisture Point	28%	Transportable Moisture Limit		25%	Moisture Content Cargo	23%
Density of Cargo [t/m³]	Angle of Repose [°]	Section Length [m]	Section breadth [m]	Angle α	Displacement [mt]	No. Of Hatches
2,30	31,0	32,00	30,00	4,0	56000,000	5

Horizontal shifting moment	30365,40
Vertical Shifting Moments	30365,40
Rise of KG	0,542
Old KG	9,5
New KG	10,042
KM prior shifting	10,73
Final GM	0,69

Final List after cargo shifted[°]	17,7

Deck edge immersion in still water	16
Allowed increase: 65%	10,4
New deck edge Immersion	26,4
Consequence if Cargo shifted	
Risk of Capsizing	
Pre-requisite Ratio TML/MC	
Cargo can be transported	
Accept Cargo	

The final list is 17,7°, which is above the deck edge immersion in still water but not above the new deck edge immersion (should not exceed 65% of the deck edge immersion in still water). Therefore a risk of capsizing must be considered and the correct course of action to be considered.

The GM, compared to the first example is also drastically decreased

Example III - Angle of natural repose = 29°

Calculation of Shifting Moments and Rise of KG for shifted Mineral Concentrates

Flow Moisture Point	28%	Transportable Moisture Limit		25%	Moisture Content Cargo	23%
Density of Cargo [t/m³]	Angle of Repose [°]	Section Length [m]	Section breadth [m]	Angle α	Displacement [mt]	No. Of Hatches
2,30	29,0	32,00	30,00	6,0	56000,000	5

Horizontal shifting moment	68601,25
Vertical Shifting Moments	68601,25
Rise of KG	1,225
Old KG	9,5
New KG	10,725
KM prior shifting	10,73
Final GM	0,00

Final List after cargo shifted [°]	35,8
Deck edge immersion in still water	16
Allowed increase: 65%	10,4
New deck edge Immersion	26,4
Consequence if Cargo shifted	Capsizing
Pre-requisite Ratio TML/MC	Cargo can be transported
	Accept Cargo

The final list after cargo shifted is far above the deck edge immersion in still water and the allowed 65% exceeding of the deck edge immersion in still water. The consequence is "Capsizing of the ship". The GM equals zero, the ship has no up-righting moments anymore.

If we would check the statical and dynamic stability curve, we could learn from the curve that the residual area is also no longer in the limits mandatory required and the angle of down flooding has also decreased, due to permanent progressive down flooding if reaching this angle.

8.7.6 Conclusion

> As larger the sections will be, as larger the result of the integral, because if solving the integral the Nominator will also increase ($L*b^4$) at a constant denominator (=4).

> Further with an increase of the angle α (35° - natural angle of repose of the cargo) means with an decrease of the angle of repose, the result of the horizontal and vertical shifting moments will increase which will also result in an increase of the rise of the KG, and therefore to a large reduction of the GM. The GM can be reduced up to 80 - 100 cm.

> The results of the dynamic stability curve are accordingly. The dynamic stability is reduced and therefore the residual area.

By drawing the statical stability curve and calculating the dynamic stability under the curve, the new found results will represent a drastically change and a non - complying with the stability regulation.

> If the ship heels more and more the ship will reach his angle of loll and beyond. This will lead to a capsizing of the ship. The time from shifting of cargo, due to liquefaction, until capsizing is sometimes very short.

Further the angle of deck edge immersion will increase. The angle should not increase by 65% of deck edge immersion in still water condition.

The calculation above shows that the master and the ship owner should carefully check and decide if Iron ore concentrates can be loaded. The prerequisite is that all necessary information are on hand and valid.

Masters and Officers can recognize if such a situation occur. For example a constant heeling to one side and the ship will not come back to its original position or comes back very slowly.

Masters and Officer should calculate the transportable moisture limit and compare it to the moisture content. Special and additional checks to be carried if it was raining and the pile is uncovered. Here the moisture concentration is increasing. New laboratory test to be conducted.

8.8 Direct Reduced Iron Ore

Direct-reduced iron (DRI), also called sponge iron, is produced from direct reduction of iron ore (in the form of lumps, pellets or fines) by a reducing gas produced from natural gas or coal. The reducing gas is a mixture, the majority of which is hydrogen (H2) and carbon monoxide (CO) which act as reducing agents. This process of reducing the iron ore in solid form by reducing gases is called direct reduction.

Direct Reduced Iron (DRI) is a manufactured metallic material produced by the reduction (removal of oxygen) of iron oxide at temperatures below the melting point of iron (1536° C or 2797° F). The iron oxide in either lump, concentrate, or pellet form is reduced at 800-1050°C (1472-1922 °F) by interaction with reductants (H2+CO) derived from natural gas or coal.

Typical Chemistry of DRI

Fe Total	90-94%
Fe Metallic	83-90%
Metallization	92-96%
C	1.0-2.5%
P2O5*	0.005-0.9%
S*	0.001-0.03%
Gangue *	2.8%-6%

8.8.1 The Direct Reduction

The reduction is chemical wise also called REDOX

Redox (short for **reduction–oxidation reaction**) is a chemical reaction in which the oxidation states of atoms are changed. Any such reaction involves both a reduction process and a complementary oxidation process, two key concepts involved with electron transfer processes. Redox reactions include all chemical reactions in which atoms have their oxidation state changed; in general, Redox reactions involve the transfer of electrons between chemical species. The chemical species from which the electron is stripped is said to have been oxidized, while the chemical species to which the electron is added is said to have been reduced. It can be explained in simple terms:

- **Oxidation** is the *loss* of electrons or an *increase* in oxidation state by a molecule, atom, or ion.
- **Reduction** is the *gain* of electrons or a *decrease* in oxidation state by a molecule, atom, or ion

8.8.2 Reducing Iron Ore – Direct Reduction

a. <u>With H_2</u>

$$3Fe_2 + O_3 + H_2 = 2Fe_3 + O_4 + H_2O$$

$$Fe_3 + O_4 + H_2 = 3FeO + H_2O$$

$$FeO + H_2 = Fe + H_2O$$

b. **With CO**

$$3Fe_2 + O_3 + CO = 2Fe_3 + O_4 + CO_2$$

$$Fe_3 + O_4 + CO = 3FeO + CO_2$$

$$FeO + CO = Fe + CO_2$$

The mineral concentrates or reduced /direct reduced iron ores are shipped in bulk and may liquefy if shipped at moisture content in excess of their transportable moisture limit (TML). These cargoes are non-combustible or have low fire-risks.

Consult the IMSBC Code (International Maritime Solid Bulk Cargoes Code) for further transport particulars; (Mineral concentrates; Section 4: Assessment of acceptability of consignments for safe shipment to which shippers/producers' attention should be drawn. If in any doubt Section 8: Cargoes which may liquefy (test procedures) should be consulted and on board testing carried out.

Mineral Concentrates

Mineral concentrates are the product of ore dressing operations whereby valuable metals recovered through mining operations are separated from waste rock prior to shipment to market.

In many mining operations ore is crushed and wet milled to liberate the valuable mineral. This slurry is concentrated by flotation and then filtered to form a dry mineral concentrate that is shipped to refineries to produce metallic products. The type of filtration equipment required depends upon the particle size, mineralogy and shipping requirements. As with all mining operations the required equipment is robust and designed to be reliable even under the toughest operating conditions.

8.8.3 Unprocessed Ores

According to the definition and description of unprocessed Ores, the description is (according to Wikipedia.com)

"A large range of ores are shipped in unprocessed condition. This category includes two particular type of unprocessed ores which have led to high profile casualties involving the loss of life of dozens of seafarers as a result of the cargoes liquefying on board during transit. Apart from a number of tragic losses, there is also record of several "near-misses" and time consuming charter party disputes as a result of shipping these cargoes."

Typical cargoes which might cause problems during the transportation due to liquefaction of the cargo are:

Lateritic Nickel Ore

Lateritic Nickel Ore will be shipped from Indonesia, Philippines and New Caledonia

If the moisture content is in excess of the Transportable Moisture Limit (TML) it causes the shift of the cargo and therefore serious stability problems

Properties of Lateritic Nickel Ore:

Colour:

The ore has a rusty-red colour (typical for all laterite soil types, they are inherently rich in iron and iron oxides.

The appearance:

Is that of a heterogeneous mixture of fine clay-like and larger rock-like particulars. The iron oxide and clay minerals host mineral for nickel but also for instance cobalt. The amount of nickel found in these ores is low, typically not more than 2%.

SF: Dry = 0,5 to 0.6

Humidity / Moisture: < 35% and up

Ventilation required: Yes - surface ventilation

The other category is Iron Ore Fines, which will be mainly shipped from India

8.8.4 Iron Ore Fines
According to the description found in Wikipedia:

"Iron ore is a natural material that is mined all around the world. Iron ore fines are created as a result of mining, crushing and processing the larger pieces of ore. Iron ore fines are less desirable (and of lower value) as they need to be sintered before they can be utilised, otherwise it will effectively smother the air flow in the blast furnace. The reason lump is preferred is that when it is fed into a blast furnace for steel-making, its particle size allows oxygen or air to circulate around the raw materials and melt them efficiently. Industry standards typically place iron ore lumps into the size range of 6 to 31 mm, and iron ore fines at particles of less than 6 mm. The iron content of the fines is typically above 60%."

Colour:

Grey to Rusty - red

Size:

Fine sized Material

SF: Dry = 0.45 - 0.55

Humidity / Moisture: < 10% (contract)

Ventilation: Surface Ventilation

If the moisture content is in excess of the Transportable Moisture Limit (TML) it causes the shift of the cargo and therefore serious stability problems - liquefaction of the cargo.

8.8.5 Sinter Feed

Is an iron concentrate containing fine particles and moisture. Will be loaded in Brazil

Colour:

Grey to Rusty - Red

Properties:

Free water separation resting on top of the cargo can occur - causing the shift of the cargo and therefore serious stability problems

SF: Dry = 0.4 to 0.5

Humidity / Moisture: < 10% (contract)

Ventilation: Surface Ventilation

For all of the above mentioned cargoes liquefaction of the cargo might occur. Liquefied cargo may (unexpectedly start to) flow from one side of the ship's cargo hold to the other. Therefore, due to the liquid form of the cargo a free surface effect will take place. This causes a dramatic reduce of the ship's metacentric height (GM). The effect primarily increases with the width of the hold.

Using the formula for getting the free surface moments (FSM) this relation of the width and length of the hold with regards to the free

surface moments can proof the increase of the FSM

$$FSM = 1/12 \times length \times width^3$$

The free surface moments causing an additional stability reduction which might lead to the capsizing of the vessel.

8.9 Common Hazard of Bulk Cargo

Some of the most common hazards of bulk cargoes on board ships are mentioned below:

1.0 Cargo shift:

> Cargo shift has always remained as one of the greatest dangers on bulk carriers. This problem is greater for ships carrying grain cargoes. Grain settles by about 2% of its volume.

Because of this settling, small void spaces exist on the top of grain surface. These void spaces permit the grain to shift. The free flowing characteristics of grain reduce the stability of any ship carrying it. Trimming is undertaken to reduce the danger of cargo shifting. Rolling can also cause shifting of cargo from one side to the other and reduce her positive stability resulting in the vessel to capsize.

2.0   Cargo falling from height:

Cargoes like iron ore, quartz and steel scraps are high density cargo. There is a possibility of cargo falling from height during cargo operations. Cargo may either fall from the conveyor belt of the ship loader or from the discharging grab on to the deck of the ship. People working on deck can get injured badly if hit by the sizeable lumps of the bulk cargo. It can be as bad as death. Cargo operation should always be monitored by responsible officers and care should be taken that no unwanted personnel are present on the working area of the deck. Persons who are involved in the cargo operation should wear protective clothing including hard hats, safety shoes and highly visible vests.

3.0   Dust from working cargo:

Dust is one of the most common hazards in bulk carriers. Many bulk cargoes are dusty by nature. Dust particles are small enough to be inhaled and if inhaled can have disastrous effects on health. Anyone working on the deck can be exposed to high levels of dust. Dust can cause sneezing and irritation of the eyes. Where possible it is always best to avoid exposure to cargo dust however if exposure cannot be avoided protective face masks should be worn. Those involved in cargo operation and need to be present on deck when a dusty cargo is being loaded or discharged and anyone sweeping cargo with a brush or with air should wear a suitable respirator. Filters should be renewed when soiled. Deck machinery should be properly protected as they can be adversely affected by dust.

4.0   Cargo Liquefaction:

Liquefaction is a phenomenon in which solid bulk cargoes are abruptly transformed from a solid dry state to an almost fluid state. Many common bulk cargoes such as iron ore fines, nickel ore and various mineral concentrates are examples of materials that may liquefy. Liquefaction occurs as a result of compaction of the cargo which results from engine vibrations, ship's motion and rolling and wave impact that further causes cargo agitation.

Liquefaction results in a flow state to develop. This permits the cargo to slide and shift in one direction thus creating free surface effect and reducing the GM thereby reducing stability. Shipper's declaration should be thoroughly examined by the chief officer before loading any bulk cargo. He must make sure that the moisture content of the cargo to be loaded should not exceed the transportable moisture limit to avoid liquefaction during the voyage. Often shippers' declaration turn out to be faulty. Spot checks can also be carried on board ships to check the moisture content.

5.0   Structural damage:

Heavy cargoes place high loads on the structure and structural failure is therefore probable. High density cargoes occupy a small area for a large weight that is they have a low stowage factor. It is therefore important that the tank top has sufficient strength to carry heavy cargoes like iron ore, nickel ore, bauxite etc. The load density of the tank top should never be exceeded. Tank top strength is provided in the ship's stability booklet. Exceeding the maximum permissible cargo load in any of the holds of a ship will lead to over stressing of local structure. Overloading will induce greater stresses in the double bottom, transverse bulkheads, hatch coamings, hatch covers, main frames and associated brackets of individual cargo holds. Poor distribution of and/or inadequate trimming of certain cargoes can result in excessive bending and sheer forces.

6.0   Oxygen depletion:

Sea transportation of bulk cargoes of an organic nature such as wood, paper pulp and agricultural products may result in rapid and severe oxygen depletion and formation of carbon dioxide. Thus apparently harmless cargoes may create potentially life threatening conditions. The cargo holds and communicating spaces in bulk carriers are examples of confined spaces where such toxic atmospheres may develop. Several fatal accidents can occur when people enter unventilated spaces. The IMSBC code lists the following cargoes as potentially oxygen depleting: coal, direct reduced iron, sponge iron, sulphide concentrates, ammonium nitrate based fertilizers, linted cotton seed. Various gaseous products are formed including carbon monoxide, carbon dioxide, hydrogen sulphide and hydro carbons. Entry of personnel into enclosed spaces should be permitted only when adequate ventilation and testing of the atmosphere is done with appropriate instruments. Emergency entry may be undertaken with SCBA. Some cargoes also use up oxygen within the cargo space. The main examples are rusting of steel swarf cargoes. Some grain cargoes may also deplete the oxygen content in the cargo space.

7.0 Corrosion:

Some cargoes like coal and Sulphur can cause severe damage due to corrosion. Cargoes of Sulphur in bulk are normally subjected to exposed storage and are thus subjected to inclement weather thereby resulting in the increase of moisture content of the cargo. Wet Sulphur is potentially highly corrosive. When Sulphur is loaded, any retained free water filters to the bottom of the holds during the voyage, from where it is pumped out via the bilges. Some water remains on the tank top and reacts with Sulphur. This leads to the release of sulphuric acid resulting in the corrosion of the ship's holds. Pond coal which is reclaimed after having been abandoned and dumped in fresh water ponds usually have high moisture content and Sulphur content. This type of coal may be liable to react with water and produce acids which may corrode parts of the ships

8.0　　Contamination:

Preparation of cargo holds for the next intended carriage is a critical element of bulk carrier operations. A lack of proper preparation can lead to claims related to cargo quality such as contamination, water ingress or cargo loss. Residues and dust of previous cargo can contaminate the presently loaded bulk cargo and can cause cargo stains that are not acceptable. Cement when contaminated by residues of previous cargo reduces its binding capacity. Unrefined sugar if stored near or above dry, refined sugar can damage it by the draining syrup. Water ingress may result from leaking hatch covers, back flow through bilge systems, leaking manhole lids and inadequate monitoring. Cargoes like salt can absorb moisture and dissolve into a liquid. Sugar can ferment in the presence of moisture. The bilges should be pumped out regularly during the voyage.

9.0　　Fire:

Bulk cargoes are deemed to present a great deal of fire hazards. Many bulk cargoes have a tendency to heat due to the oxidation process taking place during the voyage. Common cargoes like coal, Sulphur, cotton, fishmeal are liable to spontaneous heating. Coal also emits methane which is a flammable gas. When mixed with air it can form an explosive mixture. Dust created by certain cargoes may constitute an explosion hazard. Sulphur dust can readily ignite causing an explosion. Friction between cotton bales can cause spontaneous combustion and produce heat. Fire precautions should be strictly observed on bulk carriers.

The ship as carrier is obliged to care for the cargo in an expert manner to ensure it is discharged in the same state in which it was loaded. The IMSBC code should be consulted for the safe stowage and shipment of solid bulk cargoes. Suitable precautions and good seamanship should be adopted to minimize and overcome the hazards of bulk cargoes.

9.0 The Draught survey

Draught surveying is a commercially acceptable form of weighing that is based on Archimedes Principle, which states that anything that floats will displace an amount of the liquid it is floating in that is equal to its own weight. Briefly, the weight of the ship is determined both before and after loading and allowances made for differences in ballast water and other changeable items. The difference between these two weights is the weight of the cargo.

In order to do this the depth that the ship is floating at is assessed from the 'draught marks' and the vessels stability book is consulted to obtain the hydrostatic particulars such as the 'displacement' and other necessary data. Several corrections are required and the quantities of ballast and other consumable items need to be assessed so as to obtain the net weights as follows.

Archimedes Principle states that, when a body is wholly or partially immersed in a fluid, it appears to suffer a loss in mass equal to the mass of fluid it displaces. *Mass is the amount of matter that a body contains and is expressed in kilograms and tons. However, for the purposes of draught surveying, weight can be assumed to be the same as mass.*

If a solid block of volume 1 m3 and weight 4,000 kg is immersed in fresh water it will appear to suffer a loss in weight of 1,000 kg.

This can be verified by suspending it from a spring balance, which would indicate a weight of 3,000 kg. There is, therefore, a supporting force acting upwards that, in this case, is 1,000 kg. This is the 'buoyancy force'. The volume of water displaced by the block is obviously 1 m3, as this is the volume of the block, and 1 m3 of fresh water has a weight of 1,000 kg, and that is the buoyancy force. Therefore the buoyancy force is equal to the weight of water displaced.

Present for better explanation the illustration below

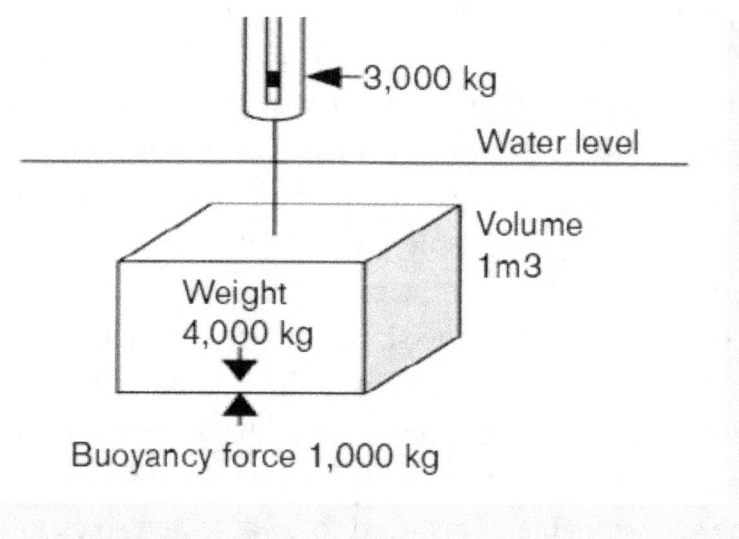

Illustration 26 Buoyancy force
(Source: Measurement of bulk cargoes Draught surveys – practice, P&I Club, Carefully to carry – May 2008)

The same solid block hollowed out, until its weight is reduced to 500 kg, and then immersed in the same fresh water will now float. This is because it still has the same volume of 1 m3 but its weight is now only 500 kg.

If the block is completely immersed, the buoyancy force will still be 1,000 kg as before, because the volume of water displaced is still the same at 1 m3. However the weight acting downwards is now only 500 kg and, once released, the block will rise until the buoyancy force acting upwards is equal to the weight acting downwards

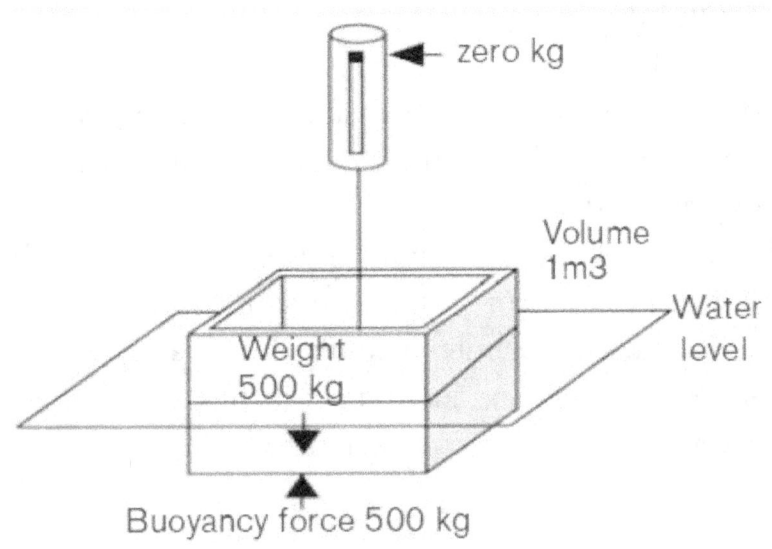

Illustration 26 (Source: Measurement of bulk cargoes Draught surveys – practice, P&I Club, Carefully to carry – May 2008)

However, had it been salt water the volume of the underwater part of the block would have been less as the density of salt water is greater than that of fresh water, meaning for equal volumes the salt water is heavier, and thus a lesser volume of it would need to have been displaced for the block to float.

From the above it can be seen that the weight of a ship can be calculated from its underwater volume and the density of the liquid in which it is floating.

In order to calculate this volume it is necessary to know how deep the ship is floating in the water as the deeper the 'draught', as it is called, the greater the weight of the ship.

Also the density of the water that the ship is floating in needs to be measured at the same time as the draughts are read.

9.1 Reading the draughts

Draught marks (the depth at which the ship is floating) are so constructed as to make the reading of them simple. Metric marks are 10 cm high and are placed 10 cm apart.

The steel plate they are made from is 2 cm wide. There are still a few ships using the 'Imperial' system but they are now few and far between. However for the sake of reference, the Imperial system has numbers that are six inches high and located six inches apart with the numbers constructed from one inch wide steel plate

9.1.1 Measuring the draughts in a swell

In turbulent conditions there may be waves, swell, pitching and rolling to take into account. In these conditions, the wave pattern should be studied to establish the wave cycle. During a series of average waves the mean of the highest and lowest draught readings should be recorded.

A total of 12 mean readings should be obtained. The highest and lowest means should be rejected and then the average of the remaining ten will give the most accurate reading possible under the circumstances.

The forward, aft and amidships draught measurements should all be found in a similar fashion.

9.1.2 Draught reading on outboard side

Every attempt should be made to read the draughts on the offshore side of the vessel although, in some situations, this may prove to be impractical or even dangerous. In such an event, the onshore marks should be read and the other side calculated with the help of a manometer.

9.2 The common method to read the draught for the draught survey

- At the time of reading the draught marks, the vessel should be upright with a minimum of trim. The trim at survey should never exceed the maximum trim for which corrections may be included in the vessel's stability book.

- Draught marks must be read on both sides of the vessel forward port and starboard; amidships port and starboard, and; aft port and starboard or, alternatively, if additional marks are displayed on large vessels at all the designated positions. Should draught marks not be in place amidships, distances from the deck line to the water line on both sides of the vessel must be measured. The amidships draughts can then be calculated from load line and freeboard data extracted from the vessel's stability booklet.

- Draught marks should be read with the observer as close to the water line as is safe and reasonably possible, in order to reduce parallax error.
- Although it is common practice to read the offside draught marks from a rope ladder, a launch or small boat provides a more stable environment and brings the observer to a safer position closer to the water line.

9.3 The general requirements for a draught survey

To calculate the cargo loaded, a draft survey must be done. Even if the ship command knows the loading rate/hr., it is quite difficult to find out how much cargo was loaded per hatch. For this reason the draft of the vessel is a nearly exact parameter for the calculation of the cargo. The draft is always corresponding to the displacement or the other way around and is corresponding to the water plane area. Draft, Displacement and water plane area have a relation to each other. If I change one of these parameters, I change the others as well. Therefore, the draft survey is a nearly exact measurement of the overall cargo loaded.

At the end of the loading, the ship command mostly will trim the ship on even keel without any list to one side. For the calculation of the cargo following figures must be on hand:

- Draft reading forward, amidships and aft – port and stb. side
- Length over the whole ship
- Length between perpendiculars (LBP)

- Forward and aft distance to perpendicular
- Longitudinal center of floatation (LCF) after loading
- Density of the water
- Bunkers, ballast, stores, freshwater and constant
- Lightship weight.

At the time of the survey the ship must be:

- Upright
- Trim as small as possible, preferable below 1% of the LBP
- No slack ballast water tanks, either full or empty
- Any ballast holds must be empty.

9.4 Calculating the deductibles

The weight of an empty bulk carrier consists of four elements
- Empty ship - this is a fixed parameter
- Stores – can be assumed as considered fixed parameter
- Ballast oil and fresh water - are changeable parameters
- Constant – is a changeable parameter

Therefore: Empty net weight = Empty ship + Stores+ Constant

The weight of a loaded bulk carrier consists of four elements
- Empty ship
- Stores
- Ballast oil and fresh water
- Cargo
- Constant

Loaded net weight = Empty ship + Stores + Cargo+ Constant
Therefore the cargo weight is the difference in the net weights.

9.5 Calculation Perpendicular corrections

Ships volumes are calculated around the section of the vessel that lies between the forward and aft perpendiculars (FP and AP). When a ship is built the draught marks are located at convenient positions on the hull and these will not always be at the perpendiculars. Therefore the draughts at the perpendiculars are required and this is done with the use of similar triangles.

The actual trim of the vessel, in relation to the length of the vessel between the draught marks, is one of a pair of similar triangles.

The next correction is in relation to the distance the draught marks are displaced from the relevant perpendicular.

Therefore these two triangles can be used to correct the draught mark readings to what they would be at the perpendiculars.

$$Fwd\ Corr = \frac{Apparent\ Trim * Fd}{LBM}$$

Where:

- ❖ Apparent trim = trim at the draught marks.
- ❖ Fd = distance of forward draught marks from Perpendicular
- ❖ LBM = length between draught marks

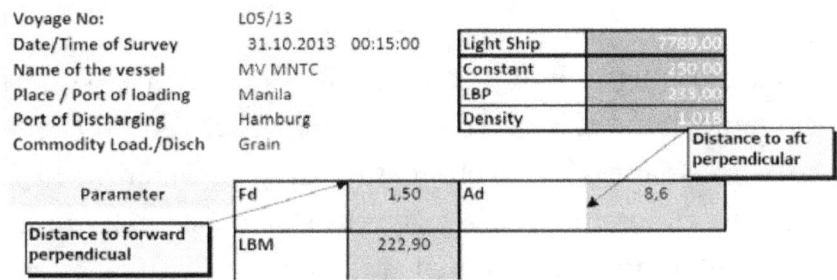

Illustration 27 Distance to Perpendicular
(Source: P.Grunau – Useful Programs – Bulk Cargo)

The calculation of the aft and amidships correction uses the same formula but substitutes the distances of the amidships or aft draught marks from the relevant perpendicular (the amidships perpendicular is located at LBP/2). See illustration above

Each of these corrections is applied according to the following rule:

If the direction of the displacement of the draught marks from the relevant perpendicular is the same as the direction of the trim, then the correction applied to the observed draught is negative, otherwise it is positive.

Simply we can conclude:

Ship is trimmed by the stern:
Fwd Correction: Minus (-); Aft Correction :Plus (+)

Ship is trimmed by the Head
Fwd Correction: Plus (+); Aft Correction: Minus (-)

If a Mid-Correction to be apply the rules remains also for the mid correction

9.6 The influence of the trim on the draught survey

A ship trims about the longitudinal center of flotation (LCF). This is the geometric center of the water plane at any time. The water plane is the area of the ship shape if it were cutoff at the water line. It obviously changes as draught increases as the shape becomes more rounded aft while remaining more pointed at the bow.
The position of LCF is important for the calculation of the draught survey. The 'true mean draught' is the draught at the LCF and is not the draught amidships. This is only the case of the LCF is at amidships

Use the illustration below for better explanation why the LCF is crucial and what is the relation between trim and LCF

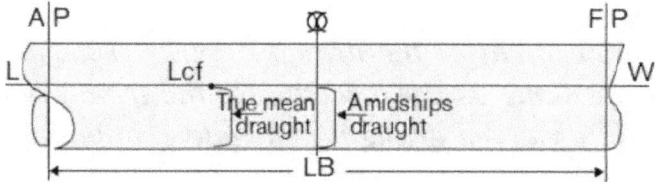

Illustration 28 Relation Trim and LCF (Source: Measurement of bulk cargoes Draught surveys – practice, P&I Club, Carefully to carry – May 2008)

Normally the LCF is located near the amidships position. If the cargo will be now loaded, the LCF is constantly changing in the direction of the highest weight distribution. If the ship has a trim by the stern, the LCF is more located aft of the amidships position, therefore the true mean draught is not amidships, because LCF is not amidships.
Therefore the trim is influencing the draught and therefore the draught survey.

9.7 Correction for deformation

The correction for deformation is known as quarter mean
Ships bend (hog or sag) due to the distribution of the
Weights. This is because of the various holds and tanks on board and not an even distribution of the cargo and deductibles. The ship is assumed to bend as a parabola and the area below a parabola, in a circumscribing rectangle. This equals to twice the area above the parabola. If we are calculating twice the area above the parabola it equals to two-thirds the total are of the parabola. The mathematics of this fact is not important from the point of view of draught surveying. What is important, is to understand the effect it has on a ship that is hogged or sagged (hogged is when the vessel is deflected upwards in its central section, and sagged is the opposite).

Explaining the deformation

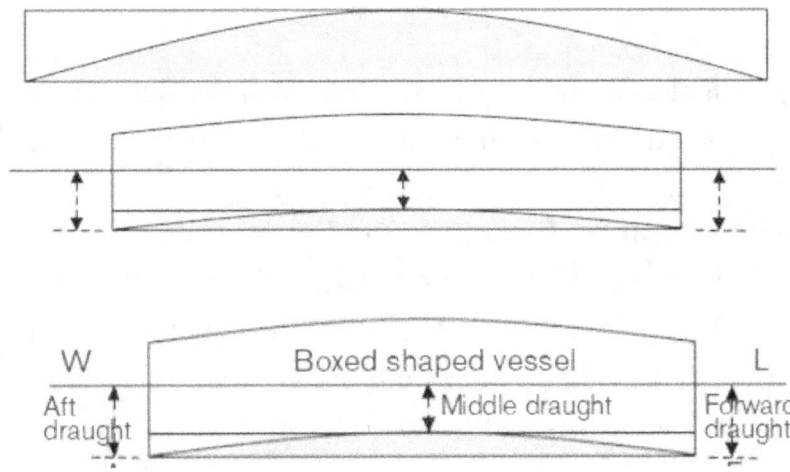

Illustration 29 Deformation
(Source: Measurement of bulk cargoes Draught surveys – practice, P&I Club, Carefully to carry – May 2008)

Utilizing the mathematics of the parabola, the lost section of volume is 2/3 of the box that encloses it. See above illustration. The formula for calculating the quarter mean is:

$$QM = \frac{(4 * Mid\ Draft) + Forward\ draft + aft\ draft}{6}$$

Is the formula presenting the lost section of volume which is 2/3?

Answer: Yes, because 2/3 = 4/6 = 66,/% of the middle draught

Calculations have shown that the most likely amount of correction required for hog or sag on a conventionally shaped ship is three quarters = 75 %.

The formula for this is called the 3/4 mean draught:

$$\frac{3}{4} Mean\ draugh = \frac{(6 * Mid\ Draft) + Fwwd\ draft + Aftdraft}{8}$$

To get the correct displacement for the quarter mean the quarter mean draught must be adjusted in accordance to the MTC. This correction is necessary for the 2nd trim correction. The procedure is simple. With the calculated quarter mean I will enter the hydrostatic table.

To get out MTC 1 and MTC 2, we are now adding 0,5 m to the quarter mean and for MTC2 we will deduct 0,5 m from the calculated quarter mean MTC must be draught. The differences between these two MTC's must be calculated and will be used later

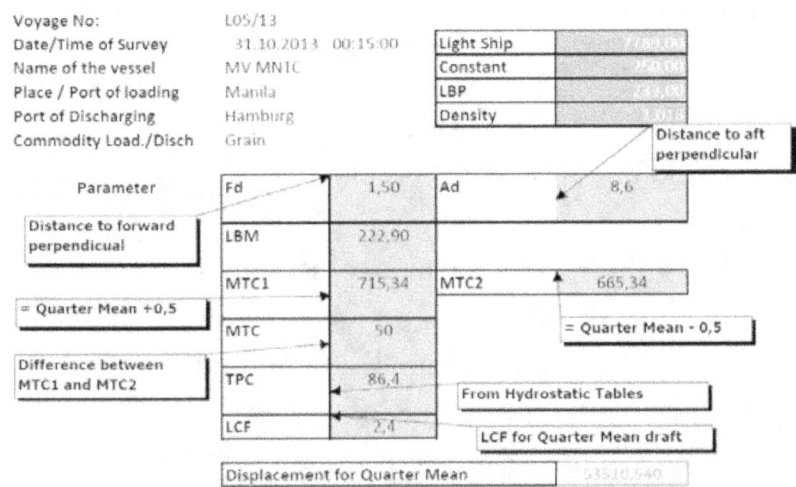

Illustration 30 Quarter Mean
(Source: Source: P.Grunau – Useful Programs – Bulk Cargo)

9.8 Correct for heel and how to apply

When a substantial heel exists, a correction for the heel must also be applied because due to the heel the water plane area is increased. **This correction is always positive.**

The effect of heel (or list) is to increase the water plane area and thus lift the ship out of the water.

Correction (in tons) = 6 x (TPC1 ~ TPC2) x (Draught1~ Draught2)

Where 1 = ship is listed to port side and 2 ship is listed to starboard side

9.9 The position of LCF

The LCF is indicated in the hydrostatic tables used for bulk carriers. There are three method how it will be indicated:
These are:
- ❖ Either with a minus (-) sign or a plus (+) sign, indicating a direction from amidships
- ❖ Or labeled with the letters 'a' or 'f' (sometimes 'aft' or 'ford') indicating aft or forward of amidships.
- ❖ Or as a distance from the aft perpendicular (in which case the distance and direction from amidships can be easily calculated by use of the LBP/2). The latter is the clearest method.

Usually the convention used is indicated at the beginning of the tables or somewhere on the pages listing the data.

LCF is the center of the of the vessel's water plane area and as such is a function of the shape of the vessel on the waterline at any given draught and nothing else.

Because the water plane changes shape to get rounder at the aft part, as the ship gets deeper, the LCF moves aft as displacement increases and forward as displacement decreases but does not necessarily move through amidships.

This means that from light to loaded condition Lcf will move either from:
- Forward to less forward.
- Forward to aft.
- Aft to more aft.

In the absence of reliable information as to the convention used in the hydrostatic tables, these facts should help to determine which side of amidships Lcf lies. Therefore, when displacement is increasing, if the actual number (indicating the position of Lcf from amidships) is decreasing, then it is forward of amidships (it is getting closer to zero, which is when it is at amidships) and if it is increasing it is aft of amidships (it has already passed zero at amidships and is moving further aft).

If on ships the LCF is not part of the hydrostatic table it can be calculated by using Simpson's Rule of approximate integration. Another method to get the LCF is to use nomograms

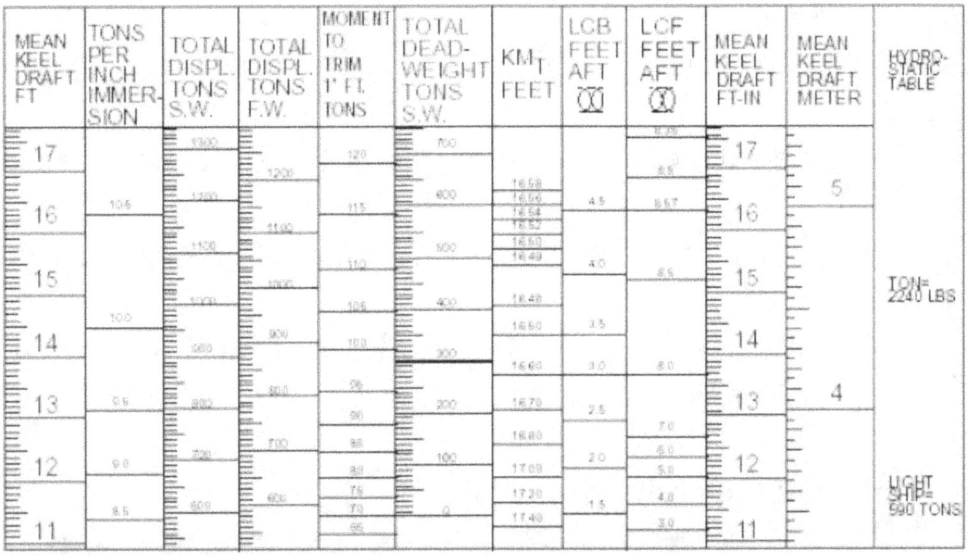

Illustration 31 Nomgram for getting LCB and LCF (Source: P.Grunau – Instructor Guide for Mathematics II)

Is it possible to calculate the LCF if the LCF for what so ever reason is not part of the hydrostatic tables of the ship?

Normally the LCF is calculated from the AP and is be its definition the centroid of the water plane at any given draught. Therefore if the water plane area changes the LCF will also change.

If the ship has a trim by the stern the LCF is also located aft of amidships, because the water plane area has changes where we have more water plane area aft of mishap.

We can conclude that with a change in Displacement and therefore with a change in draught the water plane area changes causing a change in LCF. The LCF is a function of the displacement, therefore a function of the draught and the water plane area.

For the draught survey the LCF will be normally calculated from the amidships position.

If LCF is not calculated the ship command must calculate the position of the LCF.

There are different method for getting the LCF

- ➢ Integration of the position of LCF - using Simpson's numerical integration
- ➢ Calculation of the LCF by using the trim factors fwd and aft.

Integration of the position of LCF:

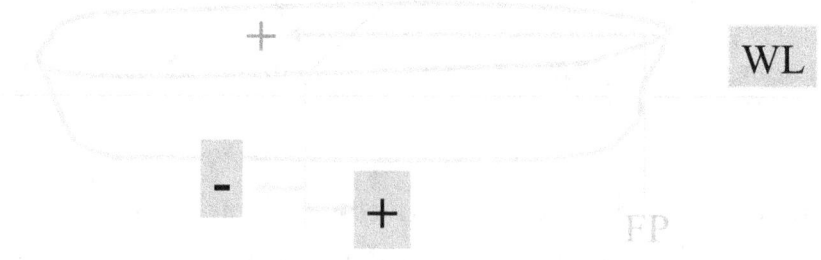

Illustration 32   Position of LCF from amidships

125

$$LCF = \int_{area} \frac{xdA}{A_{WP}} = \int_0^{Lpp} \frac{2xy(x)}{A_{WP}} dx$$

$$= \frac{2}{A_{WP}} \int_0^{Lpp} x\, y(x)\, dx$$

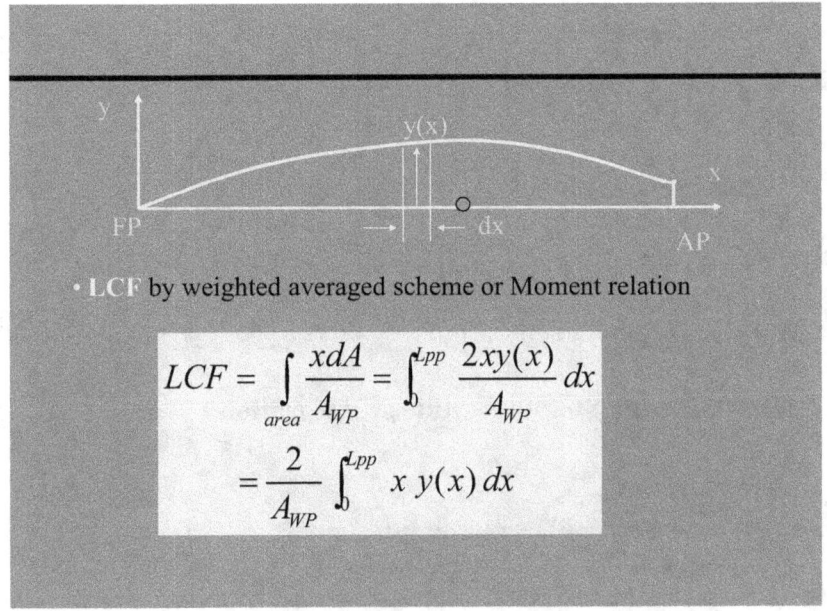

Illustration 33 Getting the Formula for integrating LCF

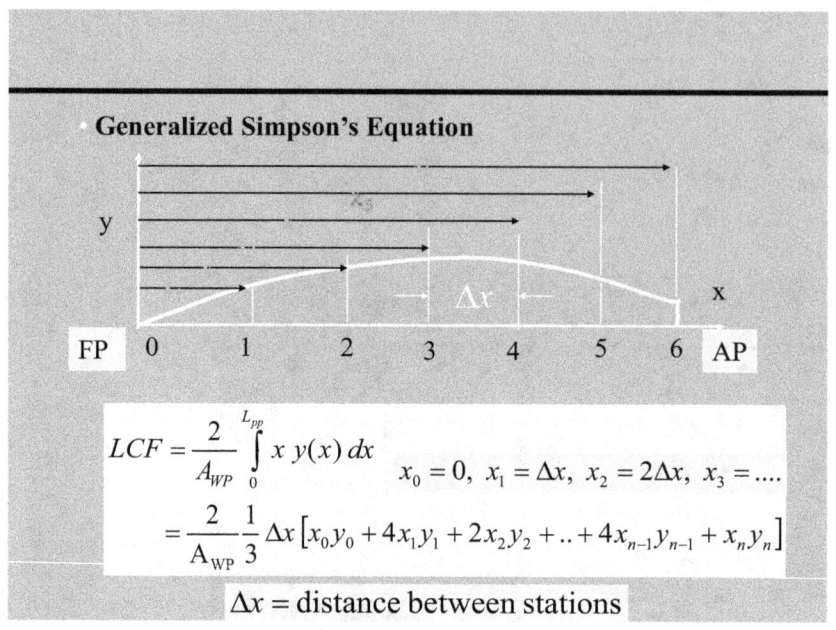

Illustration 34 Calculation using Simpson's Rule of integration

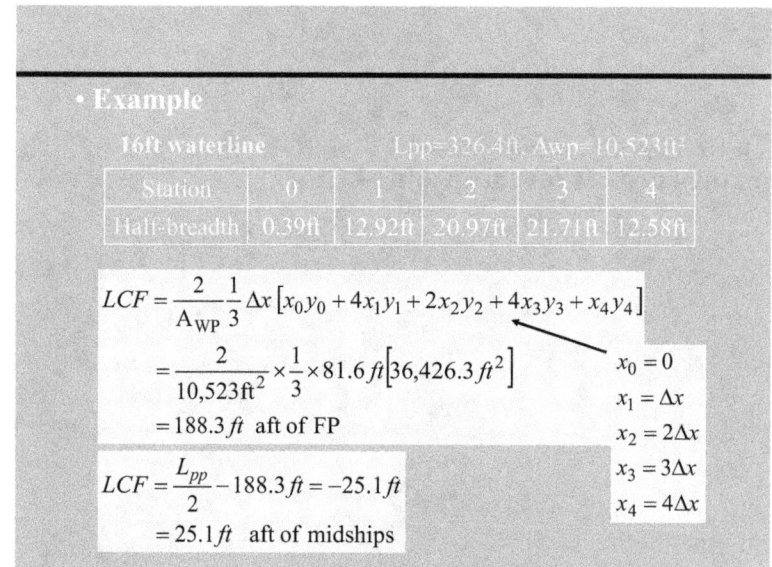

Illustration 35 Calculation example to get LCF for a certain condition
Source: PPP Lesson Mathematic – UMTC – Manila Block Training Program

The Water plane area can be either calculated using the waterline coefficient

$$WPA = L * B * Cw$$

Where L = Length of the vessel
B = Breadth of vessel
Cw = Waterline coefficient

or the water plane area can be simply estimated by using the formula

$$WPA = \frac{TPCsw * 100}{1.025}$$

Where: TPCsw is the TPC in Seawater for the actual displacement

Example:

TPC for a displacement of 57658.00 mt = 57.6 t/cm

$$WPA = \frac{57.6 t/cm * 100}{1.025} = 5619.51 m^2$$

Calculation of LCF by using the trim factors fwd and aft.

Trim factors are derived from the position of Lcf. They are a quick way for the vessel's chief mate to calculate his final trim when loading the ship.
In some hydrostatic tables the trim factors are presented but not the LCF. By using trim factors the LCF can be calculated because if the rim changes also the LCF will change

Following formula for the calculation of LCF is very common:

$$LCF\ from\ AP = \frac{LBM * aft\ trim\ factor}{fwd\ trim\ factor + aft\ trim\ factor}$$

9.10 Trim Corrections

9.10.1 1st Trim Correction

In this case the true mean draught is the draught amidships plus the layer correction. Had the LCF been forward of amidships the correction would have been negative.

The above corrections are in meters and can be applied to the 3/4 mean draughts to give the true mean draught.

However, the normal method used is to calculate the correction in tons. The displacement is taken out of the tables for the 3/4 mean draught and the layer correction applied as a negative or positive correction in tons by using the Tpc at that draught (Tpc is the number of tons required to sink the ship one centimeter).

This is the first trim correction, and is calculated using the following formula:

$$First\ Trim\ Correction = \frac{Trim[cm] * LCF[m]TPC[cm]}{LBP}$$

Here LCF is measured from amidships

Following rule can be worked out for the application of the first trim correction:

If the LCF and trim are in the same direction the correction is positive and alternatively when the LCF and the trim are in opposite direction the correction is negative

9.10.2 2nd Trim Correction

The second Trim Correction is also called Nemoto's Correction
As we already have learned, the displacement, draft and water plane area are in a direct relation to each other, meaning: the changing of one of the parameters will consequently change the other parameter as well.

The LCF data given in the hydrostatic tables is based on even keel ship. But if the ship trims, the water plane area will change shape (the LCF is the center of the water plane area and as such a function of the shape of the vessel on the waterline at any given draught). The result is that this change of shape will have an increase to the aft draft and a decrease to the forward draft or the other way around. This will also change the position of the LCF by moving further to the aft or forward to maintain the geometric position in the center. This position is normally not tabulated in the hydrostatic tables and, therefore, the second trim correction is required to compensate for this. This trim correction is named after the Japanese naval architect Nemoto. It is a compromise and is accurate up to trims of about 1% of the vessel's length.
The formula to calculate the Nemoto's trim correction equals:

$$Stc = \frac{Trim_2 * 50 * (dm \sim dz)}{LBP}$$

Where (dm~dz) is the rate of change of MTC per unit of draught – 1 meter –, it is the difference in MTC for 50 cm above and below the mean draught. The evaluation of the expression (dm~dz) is not important.

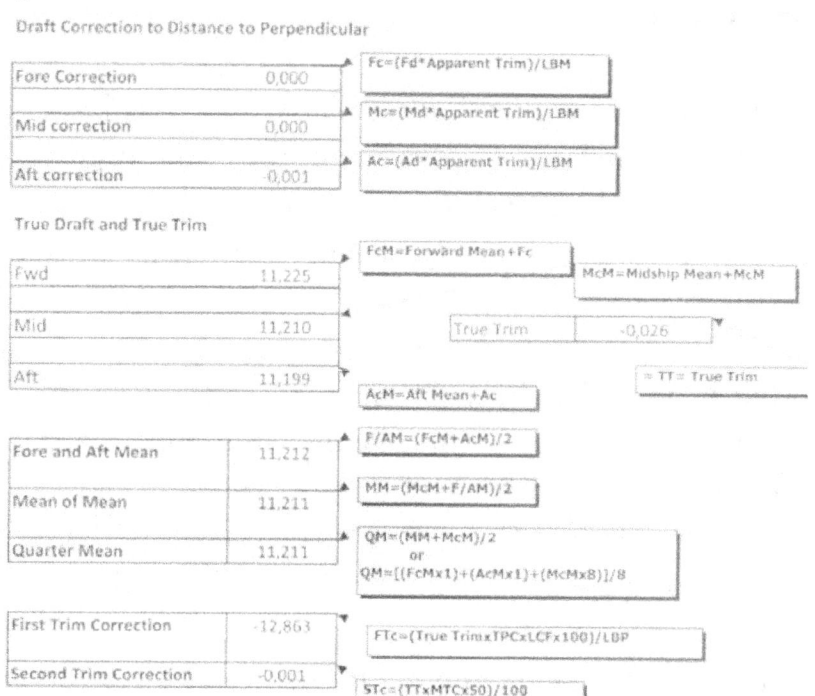

Illustration 36 Calculation of Trim Correction
(Source: P.Grunau – Useful Programs – Bulk Cargo

Some ships are supplied with 'trimmed hydrostatics'. These normally consist of several sets of hydrostatic data each one calculated for a particular trim. They may be in 10cm or 20 cm steps covering the range of trims - normally from -1 m(trim by head) to 3m trim (even keel, 1m ,2m ,3m trim by stern) over which the vessel is expected to operate. Each set of hydrostatic data consists of displacements tabulated against draught for a particular density. The densities may be in increments of 0.005 from 1.000 to 1.025 and the draughts in steps. If such tables available on board and part of the hydrostatic tables, there is no need for both the 1st and 2nd trim corrections and also the density correction.

Trimmed hydrostatic tables need to be interpolated for draught, density and trim in order to find the correct displacement. This is also the disadvantage, because the interpolation must be very exact in order to achieve an accurate result.

Another solution which might be offered and is part of the hydrostatic tables are the trim diagrams. By using the trim diagrams the trim correction is also needless. But make sure that the trim diagrams are including both corrections, the 1st and 2nd correction. Some tables are only including the 1st trim correction and not the correction for the LCF position with regards to the trim (Nemoto's Trim correction - 2nd trim correction)

9.11 The Density Correction

Once the displacement – obtained from the 3/4mean draught and the trim correction' and, if required, heel corrections – has been found it needs to be corrected for the density of the water in which the ship is floating.

The displacement of the vessel, from the ship's hydrostatic tables, is calculated at the density used to compile the tables.

When divided by this density, it gives the volume of the ship. This volume is then multiplied by the density of the water the ship is floating in to obtain the true weight of the ship.

Therefore:

a. $True\ Displ. = \dfrac{\delta sw \ast \delta dw}{\delta\ used\ to\ compilr\ the\ ship\ table}$

Where: δsw = density sea water
δdw = dock water density

Or we also can use the more common formula:

b. $Density\ correction = \dfrac{True\ Displ*(\delta actual - 1{,}025)}{1{,}025}$

The reason for saying 'density of the ships tables' in formula (a) is that some vessels are built in shipyards where 1.020 mt/m3, 1.027 mt/m3 or some other figure may be used for the hydrostatic particulars. However the norm in 99% of cases is to calculate tables at a density of 1.025 mt/m³. (Formula b)

9.12 The constant - A variable in the calculation of the Draught Survey

For calculating the estimated cargo on board, we need, as already seen, the so called "*constant*". The constant is for the ship a constant factor which must be included in the calculation, but is actually varying with the edge of the ship. There are different factors which affect the constant as there are: More stores than actually calculated, no correct bunker measuring, and no exact figures for the remaining quantity of ballast water remaining. Several layers of paint applied which will increase the ships light weight etc. According to the P&I Club – Carefully to Carry, May 2008, page 11 –
"*This stores quantity – the difference between the light ship weight and the empty ship survey – is often referred to as the 'constant'. Constant is a misnomer and it should really be referred to as a 'stores variable'.*

A ship's constant' may be affected by a variety of changes, such as under or over stated fuel figures, slops, mud in ballast tanks, incorrect ballast calibration tables, crew and stores changes, etc. and it should not be considered a fixed amount. Also a vessels light ship weight can change over the years due to a variety of additions and removals from the structure. These could be due to a variety of factors such as rebuilding, repairs, additions and modifications. As a consequence the vessels constant will include these changes unless a new light ship survey is carried out after each instance. From experience, this usually only happens after a rebuild or major additions.

The reason for a survey when the vessel is empty is to determine this variable quantity (constant). The vessel's previous experience of this constant may be the result of unreliable and badly carried out surveys. Many surveys include the lube oil in the constant and others do not. The constant can also be affected by understated fuel figures from the chief engineer, who may be keeping a quantity of oil 'up his sleeve' for a rainy day!

However, within reason this stores variable quantity (constant) can be considered to remain fixed for the duration of the ships stay in port. This is assuming that the surveyor takes note of any major changes that take place to the stores between the initial and final surveys. In other words it can be considered as a reliable measurement of the ships stores etc. for that reasonably short period of time."

All this parameters are forming the constant. For new ships the constant can be taken out of the ships manual and can be used for good.

9.13 Calculation of the displacement corrected

$$Corr.\,Displ = True\,Displ + Density\,correction$$

Where the true displacement is the displacement + the trim corrections:

$$True\,Displ. = (Displ. + (FTc + Ftc)$$

Displ.= displacement according to quarter mean.

Calculation of the cargo loaded or estimated cargo loaded:

$$Cargo\,loaded = Net\,Displ - (light\,Ship + Constant)$$

and Net Displacement equals

$$Net\,Displ. = Corr.\,Displ. - deduction(HFO, DO\,etc)$$

Use the example calculation template what you distributed before for better identifaction

Applicant

Name of the vessel
Place / Port of
Description of Cargo

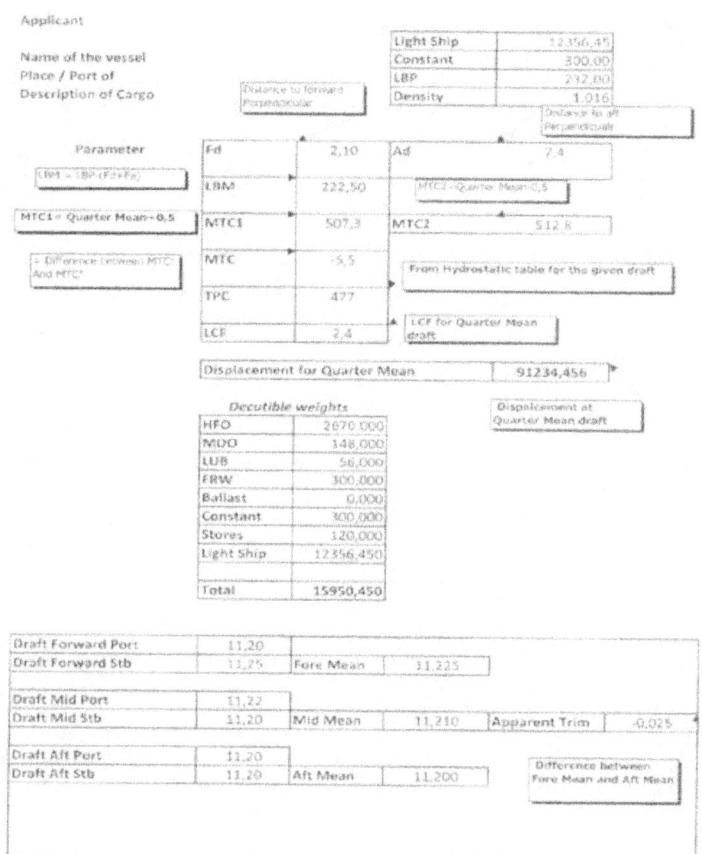

	Light Ship	12356,45
	Constant	300,00
	LBP	232,00
	Density	1,016

Parameter				
LBM = LBP-(Fd+Fa)	Fd	2,10	Ad	7,4
	LBM	222,50		
MTC1 = Quarter Mean+0,5	MTC1	507,3	MTC2	512,8
= Difference between MTC and MTC'	MTC	-5,5		
	TPC	47,7		
	LCF	2,4		
	Displacement for Quarter Mean		91234,456	

Decutible weights	
HFO	2670,000
MDO	148,000
LUB	56,000
FRW	300,000
Ballast	0,000
Constant	300,000
Stores	120,000
Light Ship	12356,450
Total	15950,450

Draft Forward Port	11,20				
Draft Forward Stb	11,25	Fore Mean	11,225		
Draft Mid Port	11,22				
Draft Mid Stb	11,20	Mid Mean	11,210	Apparent Trim	-0,025
Draft Aft Port	11,20				
Draft Aft Stb	11,20	Aft Mean	11,200	Difference between Fore Mean and Aft Mean	

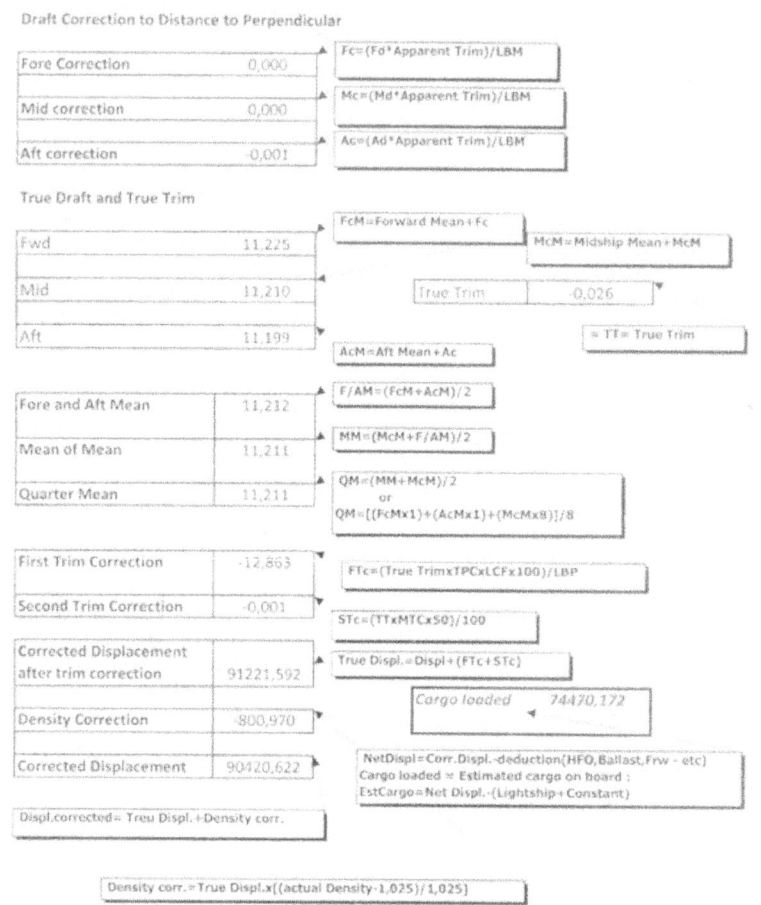

Illustration 37 Complete Draught Survey including explanations (Source: P.Grunau – Useful Programs – Bulk Cargo)

Example of a draught survey using the loading program – load master

	Port		Draughts at Marks Starboard		Mean		Port	Draughts at Perpendiculars Starboard	Mean
AFT	15.04 m	AFT		15.06 m	15.05 m	Aft:	15.04 m	15.06 m	15.05 m
MID	15.18 m	MID		15.17 m	15.18 m	Mid:	15.18 m	15.18 m	15.18 m
FORE	15.28 m	FORE		15.31 m	15.30 m	Fore:	15.29 m	15.33 m	15.31 m

Draught Mean of Means: 15.18 m

Density of Seawater: 1.018 t/m3   Trim: -0.26 m   Deflection: 0.00 m   TPC: 108.41 t   LCF: 126.95 m.fr.AP.

Deductions:

waterballast	0.00 t	Displacement at 15.18 m	152593.41 t
fresh water	236.92 t	Correction due to Deflection:	15.73 t
diesel oil	300.40 t	Correction due to Trim:	-4.96 t
fuel oil	4795.35 t	Displacement at 1.025 t/m3	152604.18 t
lub. oil	0.00 t	Correction due to Density of Seawater:	-1042.17 t
other tanks	10.27 t	Actual Displacement:	151562.00 t
Stores /Misc.	0.00 t	Total Deductions:	29858.00 t
Light Ship	24515.05 t	Cargo Loaded in Metric Tons:	121704.00 t
		Cargo Loaded in Long Tons:	119781.99 lt

10.0 The Vetting Inspection

Is vetting compulsory?
Ship Vetting is a risk assessment process carried out by charterers and terminal operators in order to avoid making use of deficient ships or barges when goods are being transported by sea or by inland waterways.

Any ship can be involved in a maritime disaster which may result in the loss of human life, damage to the environment and infrastructure, cleanup costs, fines and loss of image for the company. Therefore, it is recommended that companies carry out a risk assessment and properly evaluate the results every time a ship is proposed for any level of business.

For many years, charterers have been entitled to vet and/ or inspect the ships they select for prospective employment prior to the conclusion of any contractual agreement with the owner under the charter party.
Any vetting (and inspection) system should of course be standardized and ensure fairness, uniformity, consistency, quality and transparency in its process of assessing the ships' suitability for carrying a particular and intended cargo in a safe and efficient manner. The charterers' vetting process must be entirely transparent to the owners under the charter party and provide them with the opportunity to rectify any deficiencies or, if necessary, to appeal the findings identified by the vetting process.
Vetting companies operate on behalf of clearly identified charterers and they should not offer vetting services which could conflict with the role of the Classification Societies.
When physical inspections of vessels are conducted on behalf of charterers, inspectors should be qualified and officially certified to

conduct such inspections. The inspection process should be standardized to ensure harmonization in the vetting of ships.

To support the SOLAS requirements, that the Master's overall responsibility for the safety of the crew, vessel and cargo, including cargo operations, should never be compromised, a terminal vetting system should also be established in order to ensure that procedures for safe and efficient cargo operations exist including, for the bulk trades, a procedure for agreement between ship and shore on safe loading and/or discharging rates for the cargo.

The OCIMF SIRE – (Oil Companies International Marine Forum, *Ship Inspection Report Programme- SIRE.* This program was originally launched in 1993 to specifically address concerns about sub-standard shipping) system is still predominantly interested in tankers of all kinds but will be used in the dry bulk segment as well and is now incorporating all segments of shipping

11. Preparation of Hatches if loading Bulk Cargoes

11.1 General preparation

Before loading a bulk cargo, the master has usually to declare that the ship is ready to load as per the charter party requirements and charterer's and owner's instructions. Copies of the charter party should be placed onboard so that the master is able to see exactly what the ship's obligations are. The master can have this declaration accepted only when the holds have been inspected and accepted.

For this to happen, the master needs to know how clean the holds have to be to meet the charterer's requirements. This will depend on the previous cargo, the next cargo, local regulations and specific cargo interest requirements.

Whatever the previous cargo, all holds should be swept clean, and loose scale and rust removed. This is regardless if the same commodity will be loaded again

Traces of previous cargoes, such as Sulphur, Sulphur traces in coal cargoes and some fertilizer cargoes may corrode bare steel plate.

It is recommended that holds are swept clean after every cargo and the residues removed or, if reloading the same cargo type, placed to one side so that a tank top and hold inspection can be carried out.

Large amounts of cargo remaining onboard may not only cause outturn problems, but hide damage to the tank top plate.

The level of cleanliness of the hold required will vary from port to port, and shipper to shipper. As a general rule we can work out:

⇨ If nothing specific is stated, a double sweep, with a saltwater wash followed by freshwater wash, is a sensible option.

The hold cleaning requires proper planning not to delay the ship or to go off-hire, if he hatches are not proper cleaned

Illustration 38 Hold Cleaning using a Multi Jet Cleaning Equipment

Reference: The Standard – Bulk Cargoes- Hold Preparation and Cleaning – March 2011

In the dry bulk trade there are five grades of hold cleaning
1. Hospital clean, or 'stringent' cleanliness
2. Grain clean, or high cleanliness
3. Normal clean
4. Shovel clean
5. Load on top

Hospital Clean:

Hospital clean is the most stringent, requiring the holds to have 100% intact paint coatings on all surfaces, including the tank top, all ladder rungs and undersides of hatches.
The standard of hospital clean is a requirement for certain cargoes, for example kaolin/china clay, mineral sands ilmenite, fluorspar, rice in bulk, and high grades of wood pulp.

Normal Clean:

Normal clean means that the holds are swept clean, with no residues of the previous cargo, and washed down (or not, depending on charterer's requirements). This cleaning will be required if similar cargoes as the previous cargo will be loaded.

Shovel Clean:

Shovel clean means that all previous cargo that can be removed with a 'Bobcat' or a rough sweep and clean with shovels by the stevedores or crew. The master should clarify what standard is expected.

Load on Top:

Load on top means exactly what it says – the cargo is loaded on top of existing cargo residues. Usually, this means 'grab cleaned'.

This will be required by the charterer if the same commodity and grade will be loaded.

This will typically occur when a ship is employed under a Contract of Affreightment to carry, for example, a single grade of coal over a period.

With load on top, guidance may be necessary for the master on any cleaning requirements, including the use of bulldozers and cleaning gangs.

Grain Clean:

Is the most common requirement. A ship will be required to be grain clean for the majority of bulk and break bulk cargoes, such as all grains, soya meal and soya products, alumina.

The usual instructions a master of a tramping conventional bulk carrier will receive, particularly if his ship is unfixed for next employment, is "Clean to grain clean on completion of discharge"

The industry accepted definition of grain clean is provided by the National Cargo Bureau (NCB).

"Compartments are to be completely clean, dry, odor-free, and gas-free. All loose scale is to be removed."

- ⇨ all past cargo residues and any lashing materials are to be removed from the hold
- ⇨ any loose paint or rust scale must be removed

⇨ if it is necessary to wash the hold, as it generally will be, the holds must be dried after washing

⇨ the hold must be well ventilated to ensure that it is odor-free and gas-free

The definition says: any loose rust scale must be removed. What do we understand under rust scale?

Loose scale will break away when struck with a fist or when light pressure is applied with a knife blade or scraper under the edge of the scale. Oxidation rust will typically form on bare metal surfaces but will not flake off when struck or when light pressure from a knife is applied.

The United States Department of Agriculture permits a single area of loose paint or loose scale of 2.32 m², or several patches that in total do not exceed 9.26 m², before a hold is deemed to be unfit. In practice, the hold should be free of loose scale as each surveyor's interpretation of the required 'standard' may vary.

Illustration 39   Rust Scale

Prior loading of bulk cargoes and especially prior loading of grain cargoes a hatch inspection will be carried out by the surveyor. After the inspection the surveyor will issue a certificate that the hatches are ready to load an in an acceptable condition

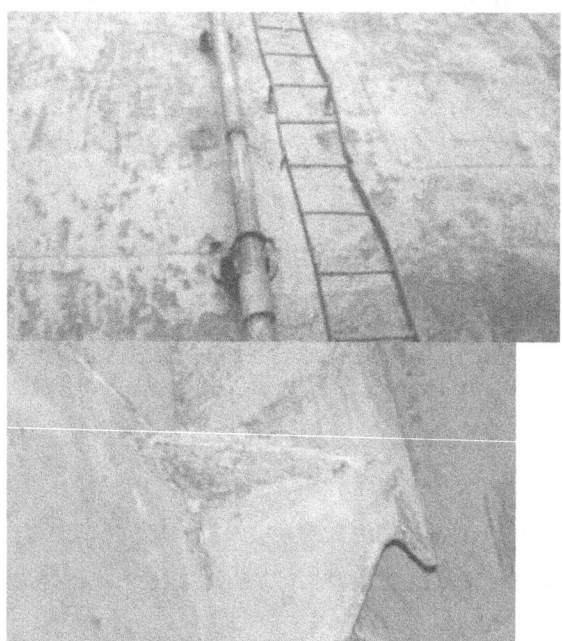

Illustration 40 Hatches failed the inspection for grain cargo

References Illustration 39 & 40: The Standard – Bulk Cargoes- Hold Preparation and Cleaning – March 2011

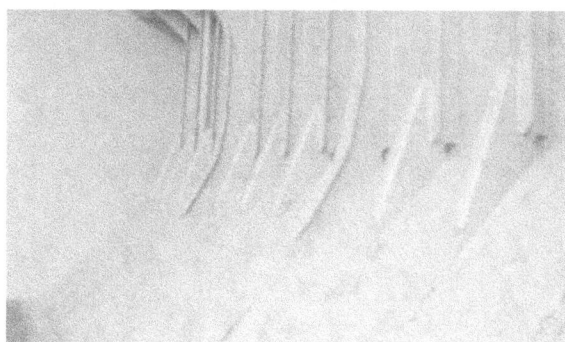

Illustration 41 Hatch accepted and ready for grain loading
Reference: The Standard – Bulk Cargoes- Hold Preparation and Cleaning – March 2011

11.2 Hold Wash

Prior hold wash, the holds must be swept and the residuals must be removed.

Afterwards the hatches must be cleaned with seawater – most common practice. Either be using the seawater hoses or be using the jet cleaning device. If using the jet cleaning device, also called the Multi jet, Combi jet or Maxi Gun, the seawater will be delivered from a hose or hoses at the pressure supplied the cleaning devices. These devices are able to deliver water – water jet – over a distance of 30 m to 40 m. The Maxi Gun can deliver a water jet of a range of 60m to 80 m.

Especial care should be taken for the residues which are behind the frames and other hatch structures or sounding pipes. During the wash down, rust scale and loose paint will be dislodged.

For some cargoes, like coal, it is sometimes advisable to use additional detergents for cleaning. These detergents will be applied as an emulsion.

A very common test for checking if the hatches are clean - USA – is the light color glove test. The surveyor is wearing a light color glove and run his hands across the bulkhead. If there will be any discoloration, the hold is not be accepted as clean and fails the survey.

After wash down with seawater the hatches must be rinsed – washed down – with freshwater to remove the salt particles.

11.3 Disposal of wash water

For the disposal of wash water the requirements of the MARPOL 73/78 Annex V must be strictly followed.

In the August 2005 amendment it is stipulated that the cargo residues are treated as garbage. Cargo residues that remain on board after discharging are thus included in the definition of garbage and need to be disposed outside special areas and as far away from the nearest land. That means: If it floats = outside of 25 nm
                  If it sinks = outside of 12 nm

Any cargo residue disposal and wash water discharge should be recorded as Garbage category 4 in the garbage record book.

For cargoes which contain oil particles, the MARPOL Annex I is mandatory. This is for disposal and discharge of wash water the same.

11.4 Bilge cleaning and preparation

After washing down the hatches with sea-and freshwater the bilges to be cleaned. Make sure that there is no residual left over, because some cargoes, like grain will create a very bad and intensive odor which will affect the next cargo.

The bilges must be absolute clean, free of any residual and the master must make sure that the bilge lines are in order and condition. Test must be carried out accordingly to justify the readiness of the bilges. All bilge alarms to be tested. (Bilge high alarm).

Following checks to be carried out:
- ⇨ Check the bilge and ballast/educator system non-return valves
- ⇨ Check the high level bilge alarms if all are operational

⇨ Check the all gasket of top tank manhole lids securely fitted with gaskets and that they are in good condition
⇨ Check the integrity of the ballast and fuel oil tanks manhole.

After completions off all bilge checks, cover the bilge plates with burlap, that grain will not block the bilges and the bilge strainer.

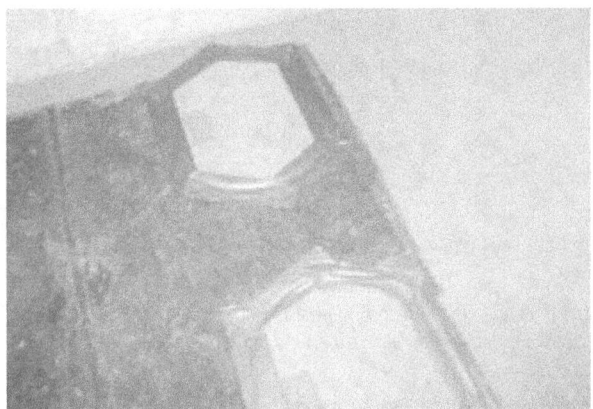

Illustration 42 Bilge plates covered and sealed, ready for loading grain
Reference: The Standard – Bulk Cargoes- Hold Preparation and Cleaning – March 2011

12.0 Hatch cover maintenance

Always

- carry out regular examination of the hatch covers, hatch beams and coamings to identify:
    - general levels of corrosion (check with your classification society for corrosion allowances);
    - localised corrosion at welded connections (grooving);
    - cracks in joints and weld metal;
    - permanent distortion of plating and stiffeners;
- call a Class Surveyor and carry out repairs as soon as possible when there are:
    - indications of excessive corrosion e.g. holes or local buckling of the top plate;
    - cracks in main structural joints;
    - areas of significant indentation, other than localised mechanical damage;
- be particularly vigilant after heavy weather;
- rectify any steel-to-steel fault before renewal of rubber packing. Renewal will not be effective if steel-to-steel contact points are defective, and expensive rubber packing will be ruined after only a few months of use;
- replace missing or damaged hatch gaskets (rubber packing) immediately. The minimum length of replaced gasket should be one metre;
- keep hatch coaming tops clean and the double drainage channels free of obstructions. (Open hatch covers to clean coaming tops and the double drainage channels after loading bulk cargo through grain or cement ports);
- keep cleats and wedges in serviceable condition and correctly adjusted;
- keep hauling wires and chains adjusted correctly;
- attach locking pins and chains to open doors and hatches;
- keep wheels, cleats, hinge pins, haul wires, and chain tension equipment well greased;
- test hydraulic oil regularly for contamination and deterioration;
- keep hydraulic systems oiltight;
- ensure the oil tank of the hydraulic system is kept filled to the operating level and with the correct oil;
- clean up oil spills. If the leak cannot be stopped immediately, construct a save-all to contain the oil and empty it regularly;
- engage tween deck hatch cover cleats when the panels are closed;
- give notice that maintenance is being performed so that no one tries to open/close the hatch;
- remember that continuing and regular maintenance of hatches is more effective and less expensive than sporadic inspection and major repair.

Reference: the Standard – A Master's Guide to Hatch Cover Maintenance

12.1 Maintenance of Hatch Cover Structure

Corrosion will reduce the strength of the hatch cover. This will lead to an increase of deflection and possible loss of the steel – to – steel contact when the cover is loaded. A good steel – to steel contact is essential and a pre-requisite for the weather-tightness of the cover. Corrosion can lead to holes in the top plate and a water/weather-tightness is no longer given.

Rubber packages should be frequently inspected and if necessary to be renewed.

The hatches are designed to drain away water that has penetrated the gasket. Therefore all draining channels should always be cleaned and free of any cargo debris. The drainage channels are located along the cross-joint and the coaming between the compression bar and the inner hatch coaming.

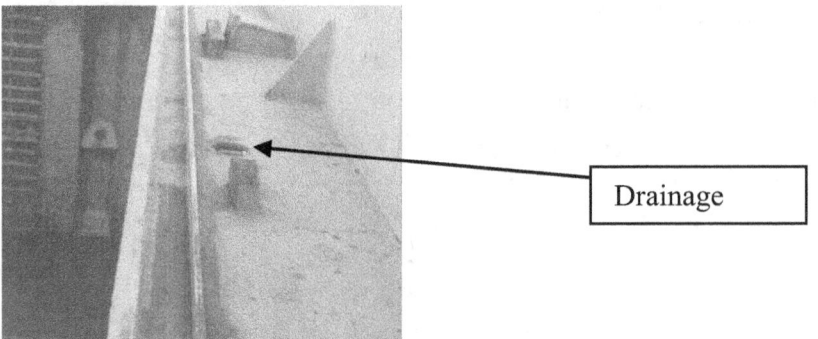

Illustration 43 Draining of Hatch – A Master's Guide to Hatch Cover Maintenance

Maintenance of hatch cleats

The hatch cleats are holding the hatch in position – gasket compression. A regular check of the washer is necessary because they are prone to physical and age hardening.

Further all hatch wedges and landing pads should be regularly inspected.

12.2 Testing of Hatch cover

Three tests are common practice now- a – days.

1. Water hose leak detection
2. Ultra-Sonic leak detection
3. Chalk testing

Water hose leak detection:

This is the most common practice on board and needs no special training. If the test is correctly performed it will show hatch joints that leak.
The general procedure is:

- Hatch cover must be closed
- Hold must be empty
- Two crew member are needed

- Take a water hose which can apply sufficient and powerful jet of water (size of the hose 25 mm – 50 mm diameter, fitted with a 12 mm diameter nozzle.
- Hold the hose 1. 50 m from the hatch joint
- Move along the joint at a speed of 1 meter every 2 seconds

Disadvantage:
The test cannot pinpoint leaks at the cross and side joints accurately.
Cannot be performed at negative atmospheric temperatures
To be avoided in ports, because the scupper plugs must be opened and that might cause pollution

Ultrasonic Leak detection test:

The ultrasonic leak detection test is an alternative test to the hose test. The ultrasonic test can be also used for leak detecting of doors and access hatches. The test is a test for weather-tightness. The advantage of this test is that the test is very accurate. Potential points can be identified.
For the test the test equipment must be class approved.
The working principle:

An electronic signal generator will be placed inside the hatch. A sensor is passed around the outside of all compression joints.
The reading taken by the senor is indicating points of low compression or of potential leakage.
The advantage of this test is that the test can be carried out if the hatches are loaded.
Disadvantages:

- The person who is carrying out the test must be an experience and trained person to interpret the readings.
- Equipment requires regular calibration
- The equipment is not available on board of all ships because it is not part of the ship's normal equipment.

Chalk Testing:

The chalk test is not a form of testing the weather-tightness of hatch covers. It will not show whether the compression is adequate.
For this test the top edge of every compression bar is covered with chalk. Afterwards the hatches will be closed. Do not use the cleats for securing the hatch.
Keep the hatch close and then reopen again. The rubber packing is examined for chalk marks. The chalk must run continuously along the packing center. If there are gaps that indicates lack of compression.
The chalk test cannot be used for testing if the hatch panels are aligned and compression is achieved.

12.3 Maintenance and Repair

Rubber gasket must be clean from any paint. If physically damages are visible the gaskets must be replaced. Check with the manufacturer's instruction.

Gasket channels are prone to corrosion. This causes the hatch packing to hang loose and to get damaged. If this is observed the gasket must be removed and the channel to be derusted and if necessary repairs to be carried out.

Compression Bars must be free of any damages otherwise the effective sealing cannot be guaranteed. If damages are visible repairs to be conducted – align the bars properly.

Landing Pads are reduced in height, repairs are essential, because the landing pads are designed to give a correct compression of the gasket when there is a metal – to –metal contact on the hatch landing pad.

Hatch Wheel Trackways are also prone for corrosion. They are weakened by abrasive wear and tear. Therefore they can distort and break. This will affect the alignment of the hatch and the movement. The hatch trackways must be always clean and painted.

Hatch cleats must be frequently adjusted. A locking nut for adjusting is located at the base of the cleat. Golden rule for adjustment: DO not overtighten the cleat

Hatch joints must be as well frequently checked. They should be always properly aligned

Drain channels and Non- return valves must be free of any loose rust or scales and residues as well. This must be checked after each preparation of the hatches for loading of the next consignment.

Greasing is absolute essential and must be conducted properly. If greasing movable parts, the parts must be moved – open and closing of hatches if wheels will be greased. Greasing of tooth rack and cylinder spherical bearings is essential. Greasing should be conducted every month.

13.0 The Grain Code and the Intact Stability Requirements

The intact stability requirements that must be met throughout the voyage (loading, discharging and the voyage) for ships carrying grain cargoes in bulk are issued in the document of authorization. These regulations are:

1. The angle of heel due to shift of cargo shall always be below 12°

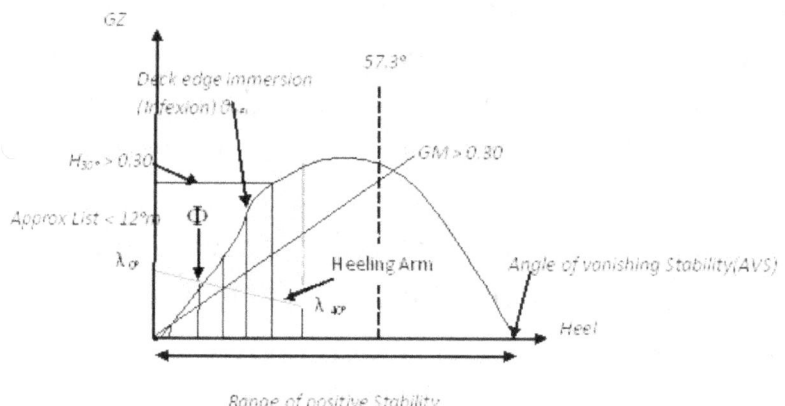

Illustration 44 Stability Criteria for Grain Cargo
Source: P. Grunau: Stability for Deck Cadets – A Working Script for Deck Cadets of the 2nd and 3rd Block

2. In the static-stability diagram, the residual area between the heeling arm curve and the GZ (righting arm) curve up to the angle of heel of maximum difference between the ordinates of the two curves, or 40°, or the angle of flooding, whichever is the least, shall in all conditions of loading be less than 0,075 meter –radians

3. The initial GM, after correction for the free surfaces shall not be less than 0, 30 m or that GM given by the formula, whichever is the greater.

$$GM_R = \frac{LxBxVd(0{,}25B - 0{,}645\sqrt{VdB})}{SFx\Delta x 0{,}0875};$$

Where L = total combined length of all full compartments [m]

B = moulded breadth of the vessel [m]
SF = Stowage factor [m³ / t]
Vd = calculated average void depth calculated in accordance with regulation B1 [m]

4. Before loading, the master shall demonstrate to the authority or the contracting government of the country of the POL the ship's ability to load and comply with the stability regulations of the Grain-Code at all stages of the voyage.

5. After completion of loading, the master shall ensure that the ship is upright before proceeding to sea.

13.1 Stability Regulations and Requirements when Loading Grain Products

⇨ It has to be assured that the cargo hold will be nearly filled.

⇨ If hatches are only partly filled, the cargo must be trimmed even.

⇨ If necessary, a center bulkhead is to be set (longitudinal or transverse), so that the cargo will not shift (shifting boards can be used, too).

⇨ If necessary, the unfilled spaces are to be filled up with bags of grain until the required stowage height is reached.

⇨ Due to the fact that these calculations will exceed the normal procedure on board, the classification societies require fixed grain loading plans.

These loading plans must contain:
- Table of volumetric heeling moment for the possibility that the grain will shift in full holds, or combination of common loading (tween decks and upper decks).

- Diagrams of volumetric heeling moments, cargo volume and center of gravity in partly filled holds according to the filling height.

- The resulting grain surface after shifting, which is assumed at 15° to the horizontal for filled compartments, trimmed grain.

- For filled compartments, untrimmed, the resulting grain surface after shifting assumed to be at an angle of 25° to the horizontal (see Illustration 77).

- Calculated cargo loading examples with different stowage factors.

The validation and approval of these plans and documentation must be presented from the ship command to the port authorities if requested. The cargo calculation must be done for the worst expected condition during the voyage. For small vessels (under 500 GRT) the requirement of calculation regarding SOLAS is not valid.

Illustration 45 International Grain Code Regulation B 2.3 & 5.1

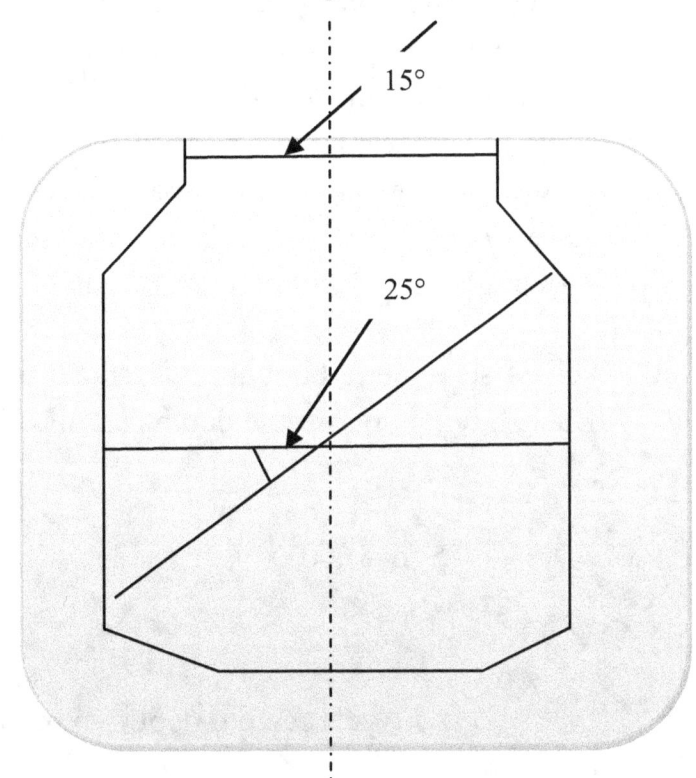

Source: P.Grunau

13.2 Voids in Spaces Loaded with Grain

In general, we are always dividing the loaded cases in: filled compartments untrimmed and trimmed.

Filled compartment, untrimmed: If a compartment is filled, untrimmed, the cargo space is filled to the maximum extent possible in way of hatch opening but has not been trimmed outside the periphery of the hatch opening.

Filled compartment, trimmed: Refers to any cargo space in which, after loading and trimming, the bulk cargo is at its highest possible level.

For the calculation of the heeling moment due to shift of cargo surface it is to be assumed that in filled compartments which have been trimmed a void exists under all boundary surfaces having an inclination due to the horizontal less than 30°.

In filled compartments, untrimmed, it is be to assumed that the surface of the grain after loading will slope in all directions away from the filling area at an angle of 30° from the lower edge of the hatch beam.

The void which exists under all boundary surfaces is parallel to the boundary surface, having an average depth calculated according to the formula:

$$V_d = V_{d1} + 0,75\,(d - 600)$$

V_d : Average void depth, in mm (see table)

Where:

V_{d1} : Standard void depth, in mm,
d : Actual girder depth, in mm.

In any case, V_d is to be assumed equal to or greater than 100 mm.

Distance, in m, from hatch end or hatch side to boundary of compartment	Standard void depth V_{d1}, in mm
0,5	570
1,0	530
1,5	500
2,0	480
2,5	450
3,0	440
3,5	430
4,0	430
4,5	430
5,0	430
5,5	450
6,0	470
6,5	490
7,0	520
7,5	550
8,0	590

Illustration 46 Standard Void Depth [mm]

For example: The distance from hatch side to the boundary compartment after completing loading the hatch with grain equals 1,5 m and the girder depth equals 50 mm, then the average void depth is:

First we will take out of the table V_{d1} for 1, 5 m = 500 mm. Substituting the values in the formula, we will get following result:

$$V_d = V_{d1} + 0,75(d - 600) = V_d = 500 + 0,75(50 - 600) = -49,25 mm$$

If we use a distance of 8 m with the same girder depth, the result is +40, 75 mm.

13.3 The Loss of GZ if Grain Cargo Shifts

As we already have seen, the most important calculation is the calculation if the vessel is corresponding to the regulations given in the International Grain-Code with respect to the stability of the vessel. The heeling arm represents the loss of GZ at various angles of heel as a result of the assumed horizontal component of the shift of grain. If the ship is listed, the loss of GZ at any angle can be determined by:

$$\text{Loss of GZ} = GG_H * \cosine\theta.$$

θ(Theta) is the angel of heel for which the loss of GZ is calculated. The loss of GZ at 0° heel = 1,0 because the cosine of 0° = 1,0. The Grain-Code states that the statically stability diagram, the net or residual area between the heeling arm curve and the GZ curve up to the angle of heel of maximum difference between the ordinates of the two curves, or 40° or the angle of flooding, whichever is the least, shall in all conditions of loading not be less than 0,075 meter-radiant.

Illustration 47 Grain Stability Requirements

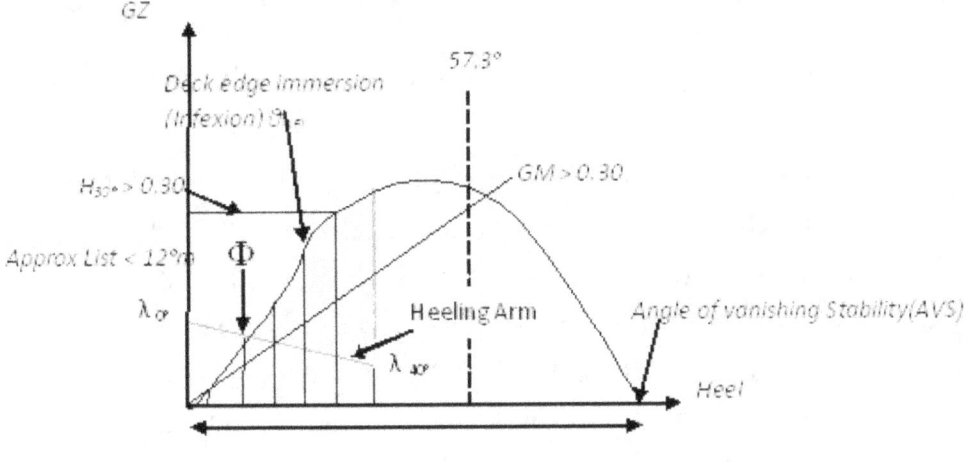

Reference: P.Grunau: Cargo handling and Stowage – BOD Verlag Norderstedt – 2015

The residual area will be limited (Illustration No.79) between the loss of GZ at 0° heel and the loss of GZ at 40° heel. For GG_H we will use now λ. Then the limitation $=\lambda_{0°}$ and $\lambda_{40°}$.

How to calculate $\lambda_{0°}$ and $\lambda_{40°}$?

As we said, $GG_H = \lambda_{0°}$, deriving for the calculation of $\lambda_{0°}$

$$GG_H = \frac{w*d}{\Delta} \therefore \lambda_{0°} = \frac{w+d}{\Delta}$$

$$weight = \frac{volume[m^3]}{SF[m^3/t]}$$

$$\therefore \lambda_{0°} = \frac{volume*d}{S.F*\Delta}; where\ volume*d = VHM[m^4]$$

For the calculation of $\lambda_{40°}$ we follow:

$$\lambda_{0°} = \frac{\Sigma VHM's}{SF * \Delta}$$ because: volume * d = VHM (see formula above)

$\lambda_{40°} = \lambda_{0°} x \cos 40° = \lambda_{40°} = \lambda_{0°} * 0,8$, because cosine 40° =0,766~0, 8

A straight line is used to join the two points of the heeling arm. The international Grain-Code ignores the fact that it must be really a cosine curve (see also Illustration No.79 – the red line within the limits $\lambda_{0°}$ and $\lambda_{40°}$).The area above this line is the residual area which should not exceed 0,075 m*rad. Where the $\lambda_{0°}$- and $\lambda_{40°}$-line intersect the curve is the maximum heel the vessel will have if grain shifts. This angle of heel must be < 12° or if the side to water is lower than the 12° maximum heel, the maximum angle of heel must be below the side to water.

Example: If the inflexion (deck edge immersion) = 10°, than the maximum angle of list should not exceed 10°, it must be < 10°. Because if the vessel has a maximum angle of heel of 12°, the vessel does already have a deck edge immersion, therefore the vessel can only heel up to maximum 9° which is below the deck edge immersion.

13.3.1 Compensation for the Vertical Component of Shift of Grain

For the calculation of the ship's KG it is common to use the height of the geometric center of the total volume in the hatch. This KG will always be higher than the actual KG of the cargo itself if any void spaces at the top of the cargo stow are also considered. As a guideline we can follow:
If the higher geometric center position will be used, the adverse effect of the vertical component of the shift of grain can be ignored.

If the lesser value for the geometric center position will be used, the volumetric heeling moments for the compartment must be increased by a factor.

The rule says:

1. In a full compartment (hatch) the volumetric heeling moments are increasing by:

    VHMs x 1,06 (see Code: Regulation B 1.3)
2. In partly filled compartments the VHMs are increasing by:

    VHMs x 1,12 (see Code: Regulation B 1.5)

The factors, 1,06 and 1,12 , are such that they will create an increase of the list of the ship which is equivalent to the effect of the vertical component of the shift of the wedge of grain (see shift as assumed by 15° or 25° for partly filled or filled hatches).

Normally, the KG of the hold, meaning the volumetric center of the hatch when the hatch is full, will be used. This means if the hatch is full, no correction is to be done. In partly filled hatches the actual grain as obtained from the hold data for partly filled hatches will be used.

Reminder: Never increase the KG only by these factors.

This calculation is necessary if correction for the VHM's are not part of the hydrostatic tables.

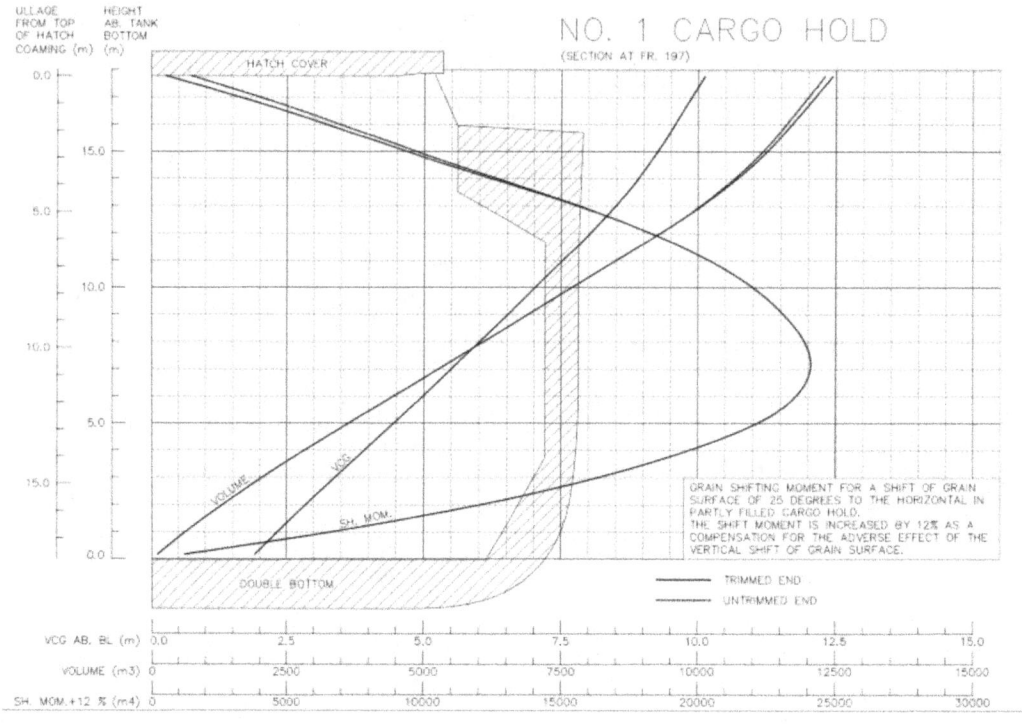

Illustration 48 Corrected Grain shifting moments
Reference: Grain Stability Booklet Diamond 53 – Handymax Bulk Carrier M/S Spar Scorpio – Chengxi Shipyard Newbuilding No 4210

In the illustration above the increase of the grain shifting moments are already included (12% increase)

HEIGHT	ULLAGE	VOLUME	SHIFT.MOM	LCG	TCG	VCG
m	m	m3	m4	m	m	m
10.775	7.200	8834.6	25097	130.02	0.00	7.38
10.975	7.000	9005.7	24624	130.02	0.00	7.48
11.175	6.800	9176.7	24137	130.02	0.00	7.58
11.375	6.600	9347.7	23635	130.02	0.00	7.68
11.575	6.400	9518.8	23118	130.02	0.00	7.78
11.775	6.200	9689.3	22586	130.02	0.00	7.88
11.975	6.000	9856.4	22048	130.02	0.00	7.98

Illustration 49: Hydrostatic Tables for shifting moments in a Hatch
Reference: Grain Stability Booklet Diamond 53 – Handymax Bulk Carrier M/S Spar Scorpio – Chengxi Shipyard Newbuilding No 4210

The grain shifting moments for the hatch should not by corrected by the factor 1.06 or 1.12. They are already corrected

Example:
Volume loaded in Hatch No 4: 9518 m³ = a filling height of 11.575m
Shifting moments: 23118 m^4

For the calculation of the actual grain heeling moments (AGHM) the volumetric heeling moments (VHM) and the compensation for the vertical component of shift of grain are needed. Further, the maximum permissible grain heeling moments must be on hand. We will get the VHM and the maximum grain heeling moments out of the hydrostatic tables. The maximum grain heeling moments are based on the actual KG and displacement after loading. The VHMs are based on each hold. The AGHMs will be calculated according to the formula:

$$AGHM'S = \frac{\sum ActualVHM'S}{SF};$$ Where actual VHMs

$= tabulated\ VHM * compensation$

If only the density is given, then the stowage factor must be calculated for this density. On most of the ships there are tables for rearranging the density into stowage factor

S.F cu.ft / tons	S.F m³ / tons	Density tone / m³
40,0	1,115	0,897
41,5	1,157	0,865
42,0	1,171	0,854
42,5	1,184	0,844

Illustration 50 Density against S.F- Source: Peter Grunau

The maximum approximate list if grain will shift should not be above 12°. The list will be calculated by comparing the actual grain heeling moments and the maximum permissible grain heeling moment. The result will be multiplied by 12° (max.12°).

$$App.List = \frac{AGHM's[tm]}{MGHM[tm]} * 12°$$; this formula can also be expressed as:

$$App.List = \frac{\frac{\sum(tab.VHM * correction)}{S.F}}{MGHM} * 12°$$

Additional to the approximate list, the loss of GZ at 0° and 40° ($\lambda_{0°}$ and $\lambda_{40°}$) must be calculated and must be entered in the statically stability curve to calculate the residual area. Instead of calculating the approximate list by means of the above formula, also the calculation of $\lambda_{0°}$ and $\lambda_{40°}$ can be done. The intersection of the Line $\lambda_{0°}$ and λ_{40} with the GZ curve represents the maximum heel if grain shifts.

Illustration 51 Calculation scheme with Explanation for Grain
Source: Grain Stability Program by Capt. Peter Grunau

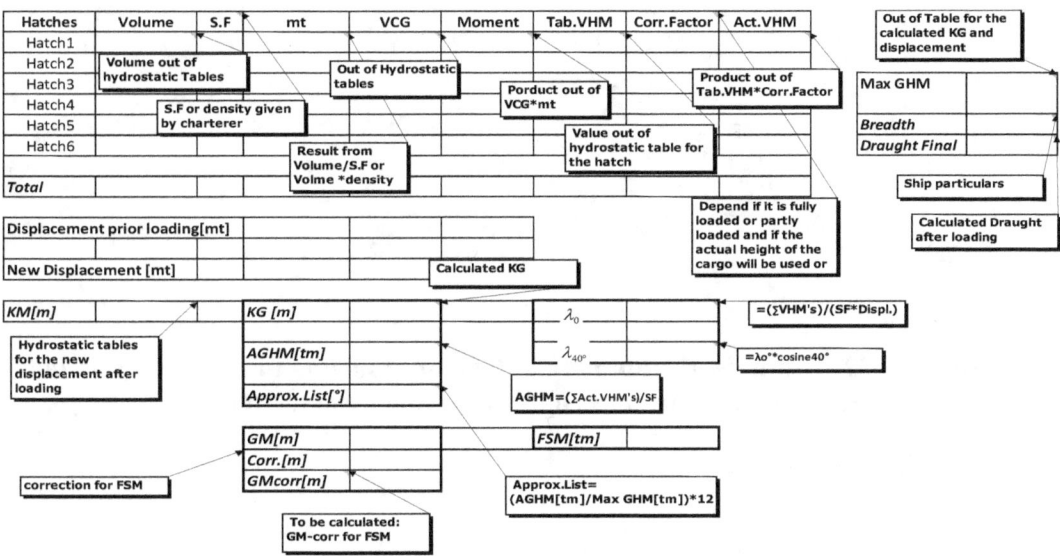

If all the requirements according to the International Grain-Code are fulfilled, the vessel can depart. Some countries have their own calculation schemes which must be followed. The ship command should clarify if the program and calculation sheet used on board can also be used or if the form to be used is the one from this authority or country.

Hatches	Volume	S.F	mt	VCG	Moment	Tab.VHM	Corr.Factor	Act.VHM
Hatch1	9076,00	1,12	8103,57	10,06	81521,93	757,00	1,06	802,42
Hatch2	12326,00	1,12	11005,36	9,41	103560,41	1034,00	1,06	1096,04
Hatch3	12570,00	1,12	11223,21	9,73	109201,88	1029,00	1,06	1090,74
Hatch4	12213,00	1,12	10904,46	9,75	106318,53	707,00	1,06	749,42
Hatch5	12310,00	1,12	10991,07	9,75	107162,95	1073,00	1,06	1137,38
Hatch6	12295,00	1,12	10977,68	9,38	102970,63	1034,00	1,06	1096,04
Total	70790,00		63205,36		610736,31			5972,04

Max GHM	6340,00
Breadth	32,5
Draught Final	10,8 m

Displacement prior loading[mt]	12854,67
New Displacement [mt]	76060,03

KM[m]	10,67	KG [m]	9,66	λ	0,08
		AGHM[tm]	5332,18	λ_{KP}	0,07
		Approx.List[°]	10		
		GM[m]	1,01	FSM[tm]	3450,00
		Corr.[m]	0,05		
		GMcorr[m]	0,95		

Illustration 52: Example of Volumetric Heeling Moments for Hatches Loaded with Grain – Trimmed - Source: Program Capt. Peter Grunau

13.3.3 The US Grain regulation and Requirement

Stability Calculation:

- Every vessel which loads bulk grain in the US shall submit a stability calculation for the intended voyage to the attending National Cargo Office (NCB) surveyor, before a certificate of readiness to load can be issued.
- The stability calculation must be in compliance with the provisions in the vessels loading booklet – approved by the administration of the country of registry
- The NCB form for the calculation of the stability must be used

Condition during the voyage:

The stability calculation shall show:
The departure condition and arrival condition
The condition on arrival and departing from bunkering ports
If there is any change in the ballast condition which differs from the departure condition – was not shown in the departure condition, than an additional calculation is required to show the compliance with the intact stability requirements just prior to ballasting condition. This referred to as an intermediate condition – see NCB form.

13.3.4 Calculation of the Residual Area

To calculate the residual area (which must be not less than 0.075 m-r) the area under the Gz curve from the angle of list(approximate angle of list if grain shifts or the edge to water, whatever will be less, to 40° must be first calculated. To calculate the area from the angle of list until 40° the area must be integrated.

$$Area = \int_{Angle\ of\ List}^{40°} f(x)\,dx$$

Where 40° s the upper limit and the angle of list where grain shifts is the lower limit. This can be assumed since the angle of heel at which maximum difference between the ordinates of two curves occur is greater than 40°. The angle of progressive down flooding will normally be above 40°. If the angle of progressive down flooding takes place below 40° than the angle of down flooding is the upper limit. To calculate the area it is recommended to use Simpson's first rule of approximate integration.

See also calculation template below.

The calculation sheet is based on the calculation of the differences between heeling arm curve and GZ Curve (Righting arm curve)

If Grain Products will be loaded, additional following calculation to be carried out:

$$\lambda_{0°} = \frac{\Sigma VHM's}{S.F * \Delta}$$

$$\lambda_{40°} = \lambda_{0°} * \cos 40°$$

Calculation of Rest dynamic Stability for grain if cargo shifts

$$A_{Rest} = \Delta\Phi * \frac{\pi}{180°} * \frac{1}{3}(4y_1 + 2y_2 + 4y_3 + y_4); \text{ where } \frac{\pi}{180°} = 0.01745$$

$$Area_{ReStability} = \frac{\Sigma \text{Product} * 5°}{3} * 0.01745 > 0.075 \text{ m rad}$$

Station(Ordinate)	dx	f(y)	Simpson Mulitplier	Product
Φ			4	
2			2	
3			4	
4			1	
Total				

Area$_{\Phi-40°}$ = [   ]

Illustration 53 Calculation form for getting the residual area
Reference: P.Grunau: Cargo handling and Stowage – BOD Verlag Norderstedt – 2015

There are two possible solutions for solving the area under the curve for getting the residual area.

Method 1 – Using the total area under the curve and subtracting the area under the trapezium

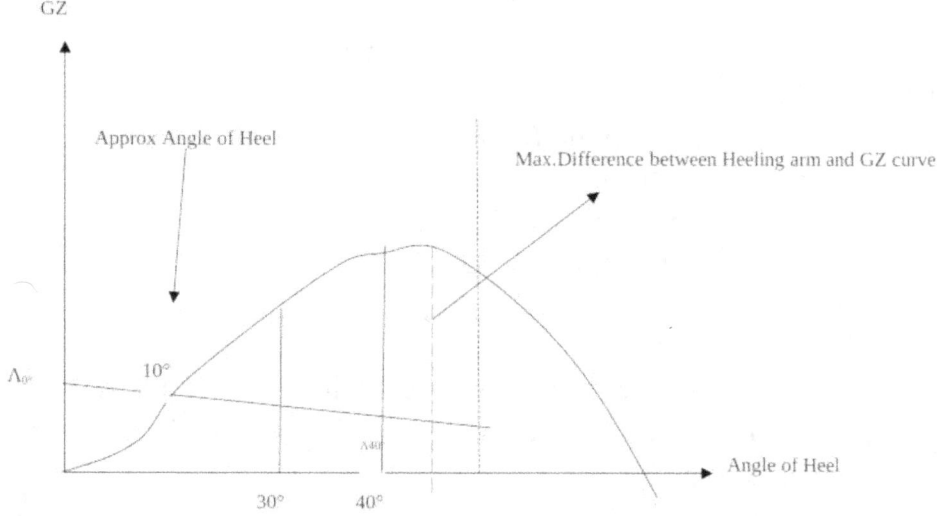

Illustration 54 GZ and Heeling Arm Curve– Source: P.Grunau

1. The max. angle which should be used is either 40° or the angle of down flooding, if the angle is lesser than 40° or the angle at which there is a maximum difference between righting arm curve and heeling arm curve. Whichever is least should be used as the limiting angle

2. Calculate the CI(common interval) to use Simpson's rule of approximate integration

$$h = \frac{Limiting\ angle\ of\ Heel - Angle\ of\ Heel}{n-1}; where\ n\ is\ the\ number\ of\ ordinates$$

Example: Limiting angle of Heel = 40°
Angle of heel if grain was shifted: 10°
Ordinates from 10° to 40° = 5

$$h = \frac{40° - 10°}{5 - 1} = 7.5°$$

The CI = 7.5 °

3. Draw the ordinates in a distance from 7.5° in the static stability curve
    Here: Ordinate 0: 10°
    Ordinate 1: 17.5°
    Ordinate 2: 25°
    Ordinate 3: 32.5°
Ordinate 4: 40°

4. Where the ordinates intersecting the GZ curve, draw a perpendicular line to the GZ axis and get the GZ value for this heel accordingly.

5. Enter the values in the Simpson's calculation form

Angle	GZ	SM	Product
10°	0.12	1	0.12
17.5°	0.32	4	1.28
25°	0.54	2	1.08
32.5°	0.70	4	2.8
40°	0.82	1	0.82
		Sum	6.1

6. Get the area of the curve:

$$Area = \frac{\Sigma\, Product * CI}{3} * 0.0175 = \frac{6.10 * 7.5}{3} * 0.0175 = 0.27\, mrad$$

We have now calculated the whole area under the curve but only need the area between the heeling arm curve and the GZ curve; therefore the area under the trapezium must be calculate and subtracted from the total area to get the residual area.

7.   Calculation of the area under the trapezium

$$\text{Area under the trapezium} = \frac{a+b}{2} * base$$

Where the base in degree is divided by 57.3° to give the measure in radians

In out example:
$a = \lambda_{0°} = 0.120$ and $b = \lambda_{40°} = 0.100$
The base is from 10° until the max. difference between heeling arm curve and GZ curve = 50°.
The difference is the base in degree = 40°

Therefore

$$\frac{a+b}{2} * base * \frac{\Delta\theta}{57.3°} = \frac{0.120+0.10}{2} * \frac{40°}{57.3°} = 0.077 \; mrad$$

8.   Subtract the area under the trapezium from the total area to get the residual area

Residual area = 0.27 mrad – 0.077 mrad = 0.193 mrad.

9.   Check if the requirements given by the IMO is fulfilled. The residual area should be larger than 0.075 mrad. 0.193 mrad > 0.075 mrad. Requirement fulfilled.

Method 2: Calculate the area under the curve for getting the residual are by subtracting the distance between heeling arm curve and GZ curve.

1. The max. angle which should be used is either 40° or the angle of down flooding, if the angle is lesser than 40° or the angle at which there is a maximum difference between righting arm curve and heeling arm curve. Whichever is least should be used as the limiting angle

Illustration 55 NCB Method for getting the Residual Area

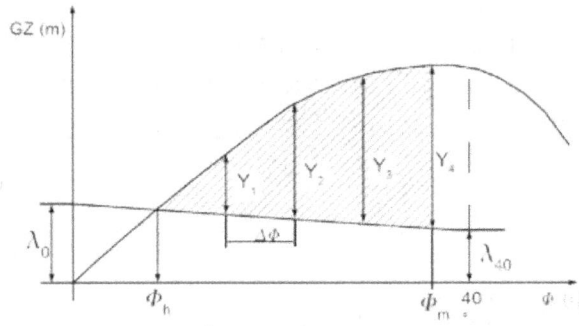

Reference: MV DortheOldendorf – Grain Loading and Stability Booklet.Compiles for training prupose by S.Wessels – Staatliche Seefahtschule Cuxhaven

1. Calculate the CI(common interval) to use Simpson's rule of approximate integration

$$h = \frac{Limiting\ angle\ of\ Heel - Angle\ of\ Heel}{n-1}; where\ n\ is\ the\ number\ of\ ordinates$$

Example: Limiting angle of Heel = 40°
Angle of heel if grain was shifted: 10°
Ordinates from 10° to 40° = 5

$$h = \frac{40° - 10°}{5 - 1} = 7.5°$$

The CI = 7.5 °

1. Calculate λ_0 and $\lambda_{40°}$

$$\lambda_{0°} = \frac{\Sigma VHM's}{S.F * \Delta}$$

For our example : VHM's = 5342 m⁴, Displ. = 30056 mt, S.F = 1.3m³/t

$$\lambda_{0°} = \frac{5342 m^4}{1.3 m^3/t * 30056 mt} = 0.137$$

$$\lambda_{40°} = 0.8 * \lambda_{0°}$$

$$\lambda_{40°} = 0.8 * 0.137 = 0.11$$

Enter the values in the graph and connect the lines. This is the heeling arm. Where the heeling arm is intersecting the first time with the GZ curve is the approximate angle of heel if grain will shift.

2. Draw the ordinates in a distance from 7.5° in the static stability curve

Here: Ordinate 0: 10°
Ordinate 1: 17.5°
Ordinate 2: 25°
Ordinate 3: 32.5°
Ordinate 4: 40°

3. Take the value form the heeling arm curve and the GZ curve and enter it in the Simpson's form.

Please be reminded that 40° is the lesser value compared to the angle of flooding and the angle of the maximum differences between heeling arm and GZ curve

Angle	GZ	Heeling Arm	Difference	SM	Product
10°	0.137	0.137	0	1	0.137
17.5°	0.32	0.13	0.19	4	0.76
25°	0.54	0.12	0.42	2	0.84
32.5°	0.70	0.115	0.585	4	2.34
40°	0.82	0.11	0.71	1	0.71
		Sum	Sum		4.787

4. Get the area under the curve:

$$Area = \frac{\Sigma Product \cdot ci}{3} = \frac{4.787 \cdot 7.5}{3} * 0.0175 = 0.209 \; mrad$$

The result represents the area under the curve. This calculation method will be also used from the US in the NCB Form.

6. Check if the requirements given by the IMO is fulfilled. The residual area should be larger than 0.075 mrad. 0.209 mrad > 0.075 mrad. Requirement fulfilled.

Example 1:

1. Angle of List if grain shift takes place is below edge to water and the angle of down flooding is above 40°
Given: Angle of List = 10°
Angle edge to water = 16°
Angle of down flooding = 55°

Therefore:
$$Area = \int_{10°}^{40°} f(x)dx$$

Calculation based on the differences between heeling arm curve and GZ curve

The common interval is now:

$$\frac{Max\, Angle - Angle\, of\, List}{n-1}; \text{ where } n = \text{number of ordinates}$$

Using our calculation sheet the number of ordinates = 4, therefore n-1 = 4-1 = 3

Getting the common interval =

$$\frac{40° - 10°}{3} = 10 = CI;$$

Let's assume the product of the area using Simpson's Rule =

Ordinates	dx	F(x) GZ	Simpson's Multiplier	Area
10	10°	0.14	4	0.56
20	10°	0.28	2	0.56
30	10°	0.55	4	2.2
40	10°	0.75	1	0.75
				4.07

Therefore: $\frac{\sum Product(simpson's) * C.I}{3} * 0.0175$

$$\frac{4.07 * 10°}{3} * 0.0175 = 0.237 m-r > 0.075\ m-r$$

The grain requirementsv are fulfilled, because the area is greater than the required minimum.

Example 2

1. Angle of List if grain shift takes place is above edge to water and the angle of down flooding is above 40°
   Given: Angle of List = 10°
   Angle edge to water = 8°
   Angle of down flooding = 55°

Calculation is based on the differences between heeling arm curve and GZ curve
Therefore:

$$Area = \int_{8°}^{40°} f(x)dx$$

$$CI = \frac{40°-8°}{3} = 10.66 = CI;$$

Ordinates	dx	F(x) GZ	Simpson's Multiplier	Area
8°	10.66°	0.11	4	0.44
18.66	10.66°	0.18	2	0.36
29.32	10.66°	0.47	4	1.88
40	10.66°	0.75	1	0.75
				3.43

$$\frac{3.43 * 10.66°}{3} * 0.0175 = 0.213 m-r > 0.075\ m-r$$

Requirement is not fulfilled even if the angel of approx. Heel if grain will shift is < 12°. But the deck edge immersion takes place at 8°. Therefore the approx. Angle of heel should not exceed 8° it should be below 8° to fulfill the requirement. Here the limiting angle is 40° because the angle of down-flooding is larger than 40° and also the maximum difference between heeling arm curve and GZ curve is larger than 40°, therefore 40° is the lesser angle.

Example 3:

1. Angle of List if grain shift takes place is above edge to water and the angle of down flooding is below 40°
   Given: Angle of List = 8°
   Angle edge to water = 10°
   Angle of down flooding = 38°

$$Area = \int_{8°}^{38°} f(x)dx$$

CI = 10°

Ordinates	dx	F(x) GZ	Simpson's Multiplier	Area
8°	10°	0.11	4	0.44
18	10°	0.18	2	0.36
28	10°	0.43	4	1.72
38	10°	0.72	1	0.72
				3.24

$$\frac{3.24 * 10°}{3} * 0.0175 = 0.189 m - r > 0.075\ m - r$$

Requirement is fulfilled because the angel of approx. Heel if grain will shift is < 12° and the deck edge immersion takes place at 8°. Therefore the approx. angle of heel is greater than angle where the deck edge immersion takes place. Here the limiting angle is 38° - angle of downflooding. For this example it is the lesser angle.

Example 4:

2. Angle of List if grain shift takes place is below edge to water and the angle of down flooding is below 40°
   Given: Angle of List = 12°
   Angle edge to water = 18°
   Angle of down flooding = 35°

$$Area = \int_{12°}^{35°} f(x)dx$$

CI = 7.66

Ordinates	dx	F(x) GZ	Simpson's Multiplier	Area
12°	7.66°	0.15	4	0.60
19.66	7.66°	0.19	2	0.38
27.32	7.66°	0.39	4	1.56
35	7.66°	0.69	1	0.69
				3.23

$$\frac{3.23 * 7.66°}{3} * 0.0175 = 0.144m - r > 0.075\, m - r$$

Requirement fulfilled. The angle of approx. heel is 12° and the area under the curve is > 0.075 mrad. The limiting angle is again the angle of down-flooding which is lesser than 40° and the maximum difference between heeling arm curve and GZ curve.

Calculation of the angle of down flooding.

The angle of down-flooding is part of the hydrostatic table. The angle will decrease with an increase of the displacement.

Further the angle of down-flooding can be also calculated. The angle of heel at which the lower edges of any openings in the hull, superstructure or deckhouse which lead below deck and cannot be closed weather tight submerge is called the **angle of down-flooding for intact stability**

The angle of down flooding can be calculated or is part of the hydrostatic tables.

Calculation :

$$\theta_f = \frac{Dist.Waterline\ to\ first\ deck\ opening}{Dist.deck\ opening\ to\ Ship\ Centerline}$$

The angle is an inversely function of the displacement and draught and therefore also a function of the WPA. If the Displacement increases, the angle will decrease.

13.3.5 Residual area according to the NCB

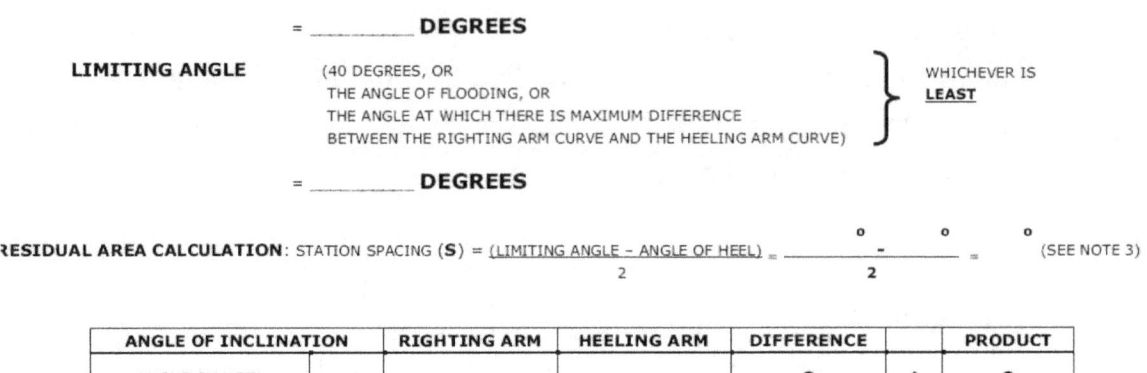

Illustration 56 NCB Calculation Scheme for getting the Residual Area
- Reference: NCB Grain stability form

The US requirements make a short cut and using only 3 ordinates. The Angle of heel or edge to water whatever is less as the low limit and 40° or if angle of down flooding is below 40° as an upper limit. There is only one ordinate in between,

therefore $CI = \frac{Max.Angle - Angle\ of\ List}{2}$; because $n-1 = 2$

This is not as accurate as the other method, because I have less ordinates. As more ordinates as more accurate.

.

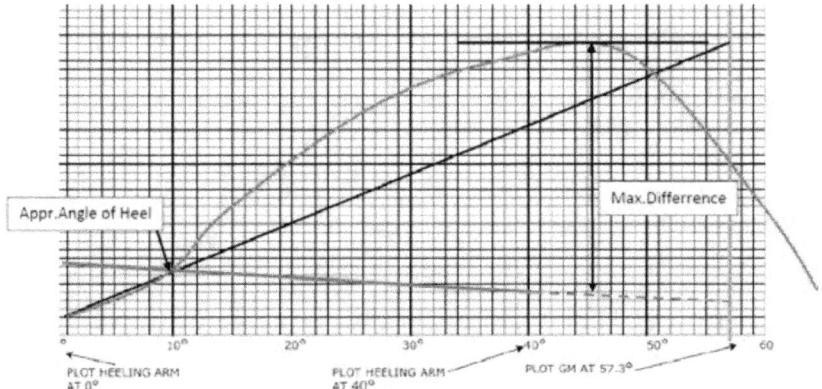

Illustration 57 Residual Area calculation
Reference: P.Grunau – own drawing

The NCB rule says:
Determine from the above plot:

a. Angle of heel (The first interaction of the righting arm curve with the heeling arm curve)
b. Limiting Angle (40°, or the angle of flooding, or the angle at which there is maximum difference between the righting arm curve and the heeling arm curve) – whatever is less

In the example the limited angle is 40° because it is less than the angle of flooding and the max. difference between the righting arm curve and heeling curve

Therefore: Station spacing = CI =
$$\frac{Limiting\ Angle - Angle\ of\ heel}{2} = \frac{40° - 10°}{2} = 15°$$

The rest of the calculation is equivalent to the calculation in the stability booklet using Simpson's first rule of approximation to calculate the residual area.

14.0 Stowage of Grain

The International Grain code is describing the stowage of grain in Bulk

> ➢ Description of sauces. If used instead of grain bulkheads

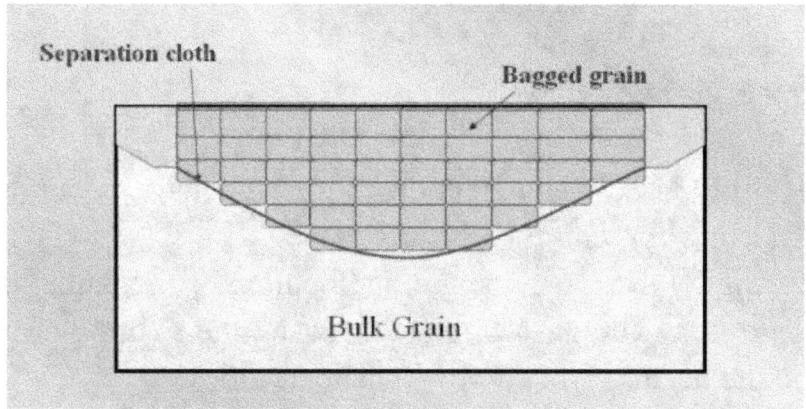

Illustration 58 Saucer
http://rswhewage.weebly.com/grain-cargo.html

Remark:
The top (mouth) of the saucer is formed by the under deck structure in the way of the hatchway, i.e., hatch side girders or coaming

The saucer and hatchway above is completely filled with bagged grain or other suitable cargo laid down on the separation cloth and stowed tightly against adjacent structures and the hatch beams

➢ Bundling of Grain Bulk

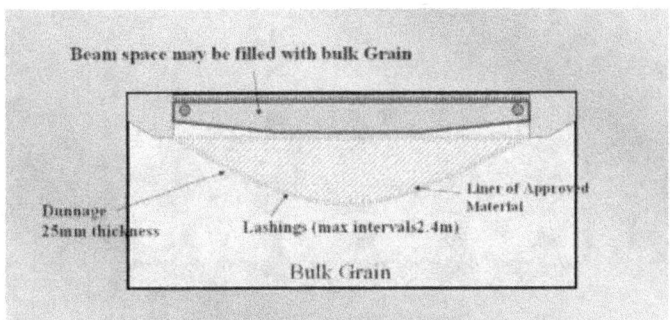

Illustration 59 Bundling of Grain Cargo
http://rswhewage.weebly.com/grain-cargo.html

Remark:
Filled compartment", shifting prevented by bundling the grain cargo. A bundle of similar bulk cargo is made by lining a saucer with tarpaulin or similar materials with suitable means of securing. Athwartship lashings to be placed inside the saucer formed in the bulk grain at interval not more than 2.4metres.

Dunnage of not less than 24mm x 150 to 300mm to be placed fore and aft over these lashings to prevent the cutting or chafing of the material which is placed thereon to line the saucer. The saucer is filled with bulk grain and secured at the top.

Further dunnage to be laid on top after lapping the material before the saucer is secured by setting up the lashings. If more than one sheet of tarpaulin is used to line the saucer, they shall be joined at the bottom either by sewing or double lap.

The top of the saucer should be made level with the bottom of the beams when these are in place and suitable general cargo or bulk grain may be placed between the beams on top of the saucer.

➢ Over- stowing of cargo

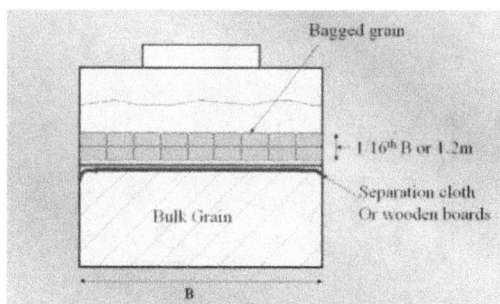

Illustration 60 Over-stowing of Grain Cargohttp://rswhewage.weebly.com/grain-cargo.html

Remark:
For a partly filled compartments –topped off by loading bagged grain or other suitable cargo

Surface to level off over and spread with separation cloth (gunny sack) or wooden boards

Over stowed with sound well filled bags to a height of 1/16th the maximum breadth of the free grain surface, or to a height of 1.2 m whichever is greater

> Shifting Boards

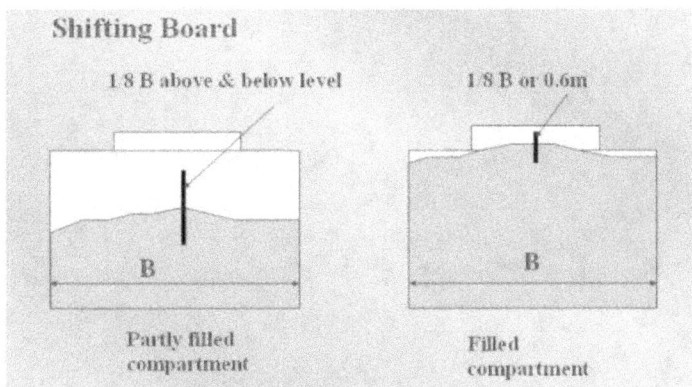

Illustration 61 Shifting Boards
http://rswhewage.weebly.com/grain-cargo.html

Remark:
Longitudinal divisions (called shifting board), which must be grain tight may be fitted in both "filled" and "partly filled compartments". In "filled compartments, they must extend downwards from the underside of the deck or hatch covers, to a distance below the deck line of at least one-eighth the breadth of the compartment, or at least 0.6m below the surface of the grain after it has been assumed to shift through an angle of 15°

In a "partly filled compartment', the division, should extend both above and below the level of grain, to a distance of one-eighth the breadth of the compartment.

Beside these topics the international grain code give guidelines and describes the calculation of heeling moments for filled and partly filled compartments due to reduction of the stability of the ship.

Bibliography

Cargo Work - Loading, Discharging&Stowage.

Club, P. (May 2008). *Measurement of Bulk cargoes.* P&I Club - Carefully to Carry.

Dibble, W., & Mitchel, P. *Draught Survey - A Guide to Good*

Practice. The North of England protection and Idemnity association Limited.

Grunau, P. (2015). *Cargo Handling and Stowage - A Guide for loading,handling,stowage, securing and transportation of different types of Cargoes, except Liquid cargoes and gas.*

Isbester, J. (1993). *Bulk Carrier Practice - A practical Guide.* Nautical Institute.

Organisation, Internationa. Maritime. *Code of uniform standards and procedures for performance of Draught Survey of Coal Cargoes.*

Standard, T. *Bulk Cargo - Hold Perparation and Cleaning*

http://de.slideshare.net/Paulo Ricardo7/draft-survey-calcualtionsheet20101.
http://www.raetsmarine.com/sites/default/files/Draft_Survey_guidelines_2010.1.pdf. ttp://www.sevensurveyor.com/draft-survey-procedures-and-calculation.

Table of Illustration

Illustration 1 Cargo Hold Structure of a single deck side skin Bulk carrier __ 5
Illustration 2 Nomenclature of typical transverse Section in way of cargo __ 6
Illustration 3 Angle of repose ___ 10
Illustration 4 Angle of repose - sheer stress and strength ___ 11
Illustration 5 Cohesiveness of particles ___ 12
Illustration 6 Multipurpose Bulk carrier, dischargingcoal in Hamburg __ 15
Illustration 7 Cargo loading plan ___ 17
Illustration 8 Discharging grain via elevators ___ 18
Illustration 9 Homogenous Loading ___ 19
Illustration 10 Alternative Hold loading condition ___ 20
Illustration 11 Block loading condition ___ 22
Illustration 12 Shearing action of the hull girder in still water ___ 23
Illustration 13 Bending of the Hull girder - sagging ___ 23
Illustration 14 Bending of the Hull girder - hogging ___ 23
Illustration 15 Permissible Stress - still water stress moments ___ 24
Illustration 16 Load planning by means of a softwarer ___ 32
Illustration 17 Loading sequence with one and two loaders ___ 39
Illustration 18 Loading Procedure if two loaders will be used ___ 40
Illustration 19: Intended Route from Norfolk to Rotterdam ___ 43
Illustration 20: Extract form Load line chart for the intended voyage ___ 44
Illustration 21 : ECA Zones of the world ___ 46
Illustration 22 Alternative Route - Southern Route ___ 47
Illustration 23 Alternative Route transferred into Load Line Chart ___ 48
Illustration 24 Grain Products and their Stowage factor ___ 70
Illustration 25 Deck edge Immersion – inflexion– ___ 89
Illustration 26 ___ 112
Illustration 27 Distance to Prependicular ___ 118
Illustration 28 Relation Trim and LCF ___ 119
Illustration 29 Deformation ___ 120
Illustration 30 Quater Mean ___ 122
Illustration 31 Nomgram for getting LCB and LCF ___ 124
Illustration 32 Position of LCF from admidships ___ 125
Illustration 33 Getting the Formula for integrating LCF ___ 126
Illustration 34 Calculation using Simpson's Rule of integration ___ 127
Illustration 35 Calcuation example to get LCF for a certain condition __ 127
Illustration 36 Calculation of Trim Correction ___ 131

Illustration 37 Complete Draught Survey including explanations_____ 137
Illustration 38 Hold Cleaning using a Mulit Jet Cleaning Equipment___ 141
Illustration 39   Rust Scale _____ 144
Illustration 40 Hatches failed the inspection for grain cargo _____ 145
Illustration 41 Hatch accepted and ready for grain loading _____ 145
Illustration 42 Bilge plates covered and sealed, ready for loading grain_ 148
Illustration 43 Draining of Hatch _____ 150
Illustration 44 Stability Criteria for Grain Cargo _____ 155
Illustration 45 International Grain Code Regulation B 2.3 & 5.1 _____ 158
Illustration 46 Standard Void Depth [mm] _____ 160
Illustration 47 Grain Stability Requirements _____ 162
Illustration 48 Corrected Grain shifting moments _____ 165
Illustration 49:Hydrostatic Tables for shifting moments in a Hatch_____ 166
Illustration 50 Density against S.F_____ 167
Illustration 51Calculation scheme with Explanation for Grain _____ 168
Illustration 52: Example of Volumetric Heeling Moments for Hatches Loaded with Grain – Trimmed _____ 169
Illustration 53 Calculation form for getting the residual area _____ 171
Illustration 54 GZ and Heeling Arm Curve _____ 172
Illustration 55 NCB Method gor getting the Residual Area_____ 175
Illustration 56 NCB Calculation Scheme for getting the Residual Area _ 183
Illustration 57 Residual Area calculation _____ 184
Illustration 58 Soucer _____ 185
Illustration 59 Bundling of Grain Cargo_____ 186
Illustration 60 Over-stowing of Grain Cargo _____ 187
Illustration 61 Shifting Boards _____ 188

Herstellung und Verlag:
BoD - Books on Demand, Norderstedt
ISBN 978-3-7412-7994-2